MW01281788

DATA PERSONIFIED

HOW FRAUD IS TRANSFORMING THE MEANING OF IDENTITY

Larry Benson and Alana Benson

Archway Publishing books may be ordered through booksellers or by contacting:

Archway Publishing
1663 Liberty Drive
Bloomington, IN 47403
www.archwaypublishing.com
1 (888) 242-5904

ISBN: 978-1-4808-6539-6 (sc)
ISBN: 978-1-4808-6538-9 (e)

Library of Congress Control Number: 2018908227

Print information available on the last page.

Archway Publishing rev. date: 08/06/2018

Contents

List of Commonly Used Abbreviations

DMV	Department of Motor Vehicles
EDRS	Electronic Death Registration System
GAO	Government Accountability Office
IRS	Internal Revenue Service
NASS	National Association of Secretaries of State
OIG	Office of the Inspector General
PII	Personal Identifying Information
SSA	Social Security Administration
TIGTA	Treasury Inspector General for Tax Administration

Foreword

My name is Frank Abagnale, but you may know me by other names I have used in the past. When you think of me, you might picture Leonardo DiCaprio, who played me in the movie *Catch Me if you Can.* If there is one thing I know, it is how to use another person's identity and defraud the system. When I boarded airplanes under the guise of a Pan-Am pilot, there was always a certain jargon I would play along with to give a convincing performance. Pilots love to talk shop, and it would be the same conversation time and time again:

"How long you been with Pan-Am?"

I'd reply, "Oh, about seven years."

"What position you fly?" The pilot would ask.

"Right seat."

"What type of equipment are you on?"

I had that one down perfectly, because whatever equipment they flew, I didn't fly. I have over 40 years of experience in the field of fraud, and one of the interesting cross-overs are these kinds of conversations, the ones that happen over and over again. The script I had memorized for flying under a fake identity may have been different than the one I used to assume an identity for a lawyer or a doctor, but they all circled around the same idea: if you talk the talk, you would be amazed at how far you can walk the walk.

While today's frauds involve fewer in-person endeavors, these conversations still happen, though now they seem to happen more online. A system can ask confirming questions like, "What's your address?" This performs a similar function as my conversation with the airline pilots. It checks me. It tries to make sure I'm in the right place, but if you're expecting the right person in the right place, you are not anticipating that they won't be. Walking through life with our eyes closed to the possibilities of fraud, we fail to notice where they are. The pilots were not looking for an imposter, so how could they find one?

Fraud operates in a similarly patterned format. In this book, there are phrases you will hear over and over, to the point of feeling repetitive. Humans love to fall into these patterns. Patterns make us feel comfortable, like we know our own terrain. And it is exactly this comfort that allows fraud to thrive. Unfortunately, the patterns we have dug our heels in over are coming back to bite us. This book explores ideas about document security and verification, communication between various agencies, and when to share data and when to keep it private. While some of it may seem like a different topic, so many of the ideas that allow fraud to happen in one avenue, are the same in another.

The connections drawn here are connections I have seen throughout my experiences with check fraud and impersonation. The lack of verification and document weakness allowed me to do everything I wanted to do. It is that same lack of verification, coupled with identification issues, that allows identity fraud to go as unchecked as it does today.

We can have these circular conversations, about how identity security is problematic, but none of it will change unless we do something to change it. We cannot count on every fraudster to have a realization in the way that I did. I was lucky to come to an understanding that what I was doing was fundamentally wrong, but not everyone has this moment. Especially now, when fraud can be perpetrated from the comfort of one's home, it is difficult to understand how your actions are affecting others, how stealing someone's identity can take away their whole life.

This is why I believe that one of the root causes of identity fraud, and in particular business fraud, comes down to one simple fact. Ethics are hard to come by. Today, few college programs offer ethical studies, and even many companies and large corporations do not have a code of ethics or code of conduct. Teaching right from wrong may sound elementary, but I have found that having a code of ethics in a company tremendously reduces the amount of fraud from employees and internal operations.

Unfortunately, the pattern of 'looking the other way' has stuck pretty well in most industries. That is why having a strong ethical code is not just the responsibility of the fraudsters, but the people trying to deter them. If you work in a DMV, and you know licenses are being fraudulently issued, do something about it. If you are a state representative, help strengthen document security. If you are a private citizen, ask questions and do not give away your information freely. We all have a responsibility, and educating ourselves on how best we can fight fraud is one of our newest civic duties.

There is no question that technology has made what I did 50 years ago much easier. In that vein, we need to finally learn the lessons we have discussed for years, and do something. At this moment, we have amazing technology that can help fight fraud and protect our infrastructure both in business and government. However, if we don't use the technology, or fail to update it, the technology becomes worthless. This book spends a fair amount of time discussing death databases. These databases are deeply flawed, and have been that way for a while. The technologies developed to ease those flaws have been poorly implemented.

Why is it that we continue to circle around issues without solving them, having conversations without discussing anything?

When I think back on the many 'fraudulent' conversations I had, the filler conversations that merely padded my fake identities, there were questions the real professionals could have asked that would have derailed my entire operation. It seems so unlikely that few people ever did ask those, but rather stuck to the mundane questions that only required mundane, easy to fake answers. If we want to adequately challenge fraud, in business, in government, or for private citizens, we need to begin asking those questions. We need to look more deeply into the causes, the lack of ethics individuals have to commit these crimes, and why our current solutions have not helped us.

Fraud is not going away. It is increasing at an exponential, and terrifying rate. I have no desire to return to the life I once led, but it would serve us well if we thought more like the fraudsters. I do not mean in the immoral way that results in stealing something from someone else, but in the way of shrewdness. I could walk onto an airplane because I escaped out of my normal pattern of 'passenger' and into the pattern of 'pilot.' I learned a different pattern, I had different conversations. The only way we can start to challenge fraudulent patterns is to learn them, infiltrate them from the inside out, and start asking the right kinds of questions.

Frank W. Abagnale

Frank Abagnale
Subject of the book, movie and Broadway musical, *Catch Me If You Can.*

Introduction

"Hello?"

"Hey, it's me. I promise it's not illegal. Can I try to get a copy of your birth certificate?"

She pauses. "Uh, I guess? What are you going to do with it?"

This is the conversation I had with my closest friend the day I asked if I could steal her identity. We have known each other our whole lives, so it probably didn't come as too much of a surprise. She did, however, ask the right question: *what are you going to do with it?*

This question is central to fraud in any form. A birth or death certificate on its own is harmless. The problem with documents is what they allow you to do, how they enable action. In many cases, possessing someone else's personal information is perfectly legal. It's normal within genealogy and ancestry studies to have other people's birth certificates lying around. Even some sports teams keep them on file. It only becomes illegal when you abuse those documents.

Our research began with a single hypothesis: it is easy to order certified birth certificates online without proving your identity. Turns out, this is true. We were shocked by how simple the process is. We spoke with CEOs of large identity information corporations, FBI officials, and government employees, none of whom were aware that an average person has the ability to order certificates containing the personal information of total strangers. Curiosity got the best of us, and we began ordering them in droves. After a few months, we had almost 30 birth certificates, none of which were our own.

We figured we would only be able to order birth certificates from what are called 'open states,' states that view birth certificates as public record, and allow anyone to order anyone else's, as long as they can provide the minimally-required, but necessary, information. This is how we ended up with Steven Spielberg's birth certificate—but I'm getting ahead of myself.

When I first started compiling our research into an outline, the hardest part was deciding on the order of what to talk about. It seemed like everything was so connected to everything else that it was impossible to create any kind of sensical order. I even dabbled with the idea of doing a 'create your own adventure format.' While I eventually gave that idea up, the premise stuck. How do you talk about documents, when to understand documents you need to understand identity fraud, which leads to business fraud because they're similar, but going back to documents they only way it makes sense is to discuss verifications, which we can't understand until we talk about death databases, which in turn, leads us back to business identity fraud.

It felt futile to come up with any logical order, so we decided to go about this book in the only way that makes sense: chronologically, in the order that we ourselves discovered the research. Every piece of information in this book connects to everything else, making it difficult to create a linear approach. Larry has worked in fraud for years. I, Alana, am newer to the fraud game, and will serve as your narrator. I want you to come to understand identity fraud as I have, discovering it and making connections as they reveal themselves. When I say 'we,' I am talking about the collaborative effort of both Larry and myself.

I started with my friend's birth certificate not just because I knew it would be easy. It was scary, taking that first tentative step into the world of identity fraud. Could I go to jail for this? Would a SWAT team descend upon my house? Of course, this was all before we figured out it was perfectly legal to order someone else's vital records. We felt a bit less hardcore after the fact.

Searching 'birth certificates' online led me to VitalChek, the website platform that carries out the specific laws surrounding access to vital records and allows people to order their certificates with ease. Started in 1987, VitalChek was the brainchild of an engineer who moved his family to Nashville, Tennessee from Delaware. After attempting to enroll his kids in school, he was informed that he needed certified copies of their birth certificates, and that the only way to obtain them would be to go in person to an issuing office in the state where they were born. The engineer flew to Delaware to obtain the certificates, and became aggravated enough by the process to start a company. VitalChek's mission became to "provide a safe, easy and convenient way obtain vital record documents—including birth certificates, marriage record, divorce record, and certain family death records."

There are several ways to order birth certificates (with variance from state to state), such as over the phone or through the mail, but ordering online is by far the easiest and least secure method. Ordering online masks the assumed identity of the person who is ordering. The orderer's identity is assumed to be true, regardless of verification.

My friend was born in Massachusetts, and when I was ordering her birth certificate, VitalChek asked me for the following information:

1. Her full name
2. Her date of birth
3. Her parents' full names (including her mother's maiden name and both middle names)
4. Her birth city
5. My relation to her, selected from a dropdown menu

I knew the answer to all these just from years of calling her on her birthday and stories about her parents' wedding. When selecting my own information, I selected 'sister'—something we've joked about enough times to feel true. Part of me worried about whether they would verify this in their own records, it would be easy enough to tell that we had no blood relation. The other part of me remembered one of the central pillars of working within bureaucracy: verification is not nearly as omnipresent as we think it is.

In this book, I will refer to it as the 'CSI Effect.' The CSI Effect is defined as, "A phenomenon reported by prosecutors who claim that television shows based on scientific crime solving have made actual jurors reluctant to vote to convict when, as is typically true, forensic evidence is neither necessary nor available."[1] While the CSI Effect has been heavily debated, the same effect seems to occur within the realm of inter-governmental communication. People en masse seem to have the impression that government agencies access some centralized database containing all personal information to ever exist. This database does not exist.

Cross-agency data sharing is rare, and verification is even more rare. How would the Vital Records agency in Massachusetts know I wasn't her sister? Perhaps she has a sister born in another state and they do not have access to that state's birth records, or maybe we did not share the same father, or maybe I was adopted. Once you begin asking questions about these systems, and their ability to communicate with each other, the holes become obvious. There is no way Massachusetts would be able to guarantee that we were not sisters, and since their specific legislation ensures blood relatives be able to order these certificates, there is no logical way for them not to grant access. I was skeptical.

These agencies have the right to retain your processing fee (generally $15-20) even if they cannot find the record you are requesting, and the website makes it explicitly clear that they will not release any record unless the information you input perfectly matches what the record says. If the father's middle name is listed as 'R.' and you list 'Randall,' in theory, you will not receive the certificate. We found this to be false on several occasions by successfully obtaining records with errors, like a missing middle name.

I typed in my friend's information, my invented sororal affiliation, and my actual name and address. I waited five days, and it showed up in my mailbox. It was laughably simple. I was unsure whether they would verify my name with her family records to see if I was who I claimed, but it did not happen. In this instance, I even wrote the wrong city of birth—and I only discovered that fact when the birth certificate itself showed up to correct me.

This was just the beginning for me. Larry had been busy ordering birth certificates for a while at this point, mostly sticking with celebrities whose personal information is as far away as a Wikipedia search. He had been focusing on the self-proclaimed open states. My friend's birth certificate was from Massachusetts, a state that, up until that point, we thought was closed.

We quickly realized that there are no such things as 'open' and 'closed' states. There is no such thing as a standard birth certificate. And there are not, that we have met, many people in the government or identity sectors who are aware of these massive holes in our identity system.

Focusing on documents and how they form the building blocks of our identities is a critical component of our research. We will also explore how identities interact with government agencies and businesses and how they grant access and enable fraud.

While we collected the majority of the data in this book ourselves, we are deeply indebted to various studies and white papers. In the first section of this book we cover the concept of state-to-state document security, and how vital records got to their current state of insane insecurity. We also look at the role of death certificates and the Social Security Administration's death data as an aid to ordering birth certificates. We discuss Social Security numbers and the key role they play in strengthening documented identities, and in the third chapter, we explore everything surrounding birth certificates: how we order them, use them, and exploit them.

The second half of this book focuses on one of the biggest and most dangerous trends in identity theft: business identity theft and fraud. By using similar tactics identity thieves use on individuals, criminals are able to walk away with bigger profits while maintaining their anonymity. Business identity fraud is taking a massive toll on small businesses and large corporations alike.

The connection between documents and business identity fraud lies in the systems used to protect their security. The data used to verify identities, whether they are individual or business identities, is often flawed and under-utilized. Improving these systems, and implementing other front-end solutions is critical to addressing fraudulent identities and protecting the real identities behind them.

Section I

Documents and Death Data

1

A History of Open States

[After Stefan tells his beloved
Elena she was adopted.]
Elena: [sob] How do you know this ?
Stefan: Your birth certificate from the city records.
—*The Vampire Diaries*, 2009

Documentation is failing to accurately serve as a checkpoint for potential fraudulent operations. There are unlimited ways for documents to aid fraudsters, and little is being done to create better failsafes for identification. We will explain some of the entry-level holes in the larger problem of systemic fraud, and how documentation and identity are at its foundation.

At this point in time, discussing documents already feels old-fashioned. The fraud game has moved online so quickly and so insidiously that it is often left out of the discussion. Understandably, it may not feel pertinent, but what documents enable individuals to accomplish only strengthens their online personas, and provide a shield behind which to hide. Even today, roughly half of identity fraud occurs without the use of a computer. Data hacks and breaches and cybersecurity, are also all critical components of fraud. Identity exists at the core of every data hack and cyber breach. If we can begin to safeguard identity from the beginning, it can perhaps alleviate some of the burden.

The concept of record keeping has existed since ancient times, from the Inca to the Babylonians to the dynasties of China. People have been keeping societal records for a very, very long time. Often, these ancient records were similar to those we keep today: goods and regional imports and exports, taxation, and census numbers. The idea of tallying these numbers is not new. Having open records, however, is a new idea. When the census records were in the hands of an autocracy or feudalistic structure, the people at the top were not so keen to allow the lower classes to see these records. For the fiefs that were literate, seeing those records would allow them to understand at exactly how outrageous of a rate they were being taxed. Here I go, explaining our tax code.

Since the early 1700s, the democratic model of the United States allowed for the development of open records. The problem began in New England where the colonists brought more than pale-faced people and disease: they brought their Western understanding of record keeping. Naturally, the need for securing birth records was by no means critical. Identity theft was not a big problem during colonial times. The tradition of the town clerk handling records in each individual town helped shape the way these records would be handled in the future.

The structure of local town clerks issuing birth certificates and other records sustained, and other strange local laws in the older states simply became outdated and forgotten. In Pennsylvania alone there are hundreds of individuals who can legally issue records. This in turn has led to thousands of different types of birth certificates, all equally valid, in circulation.

The concept of open states is tricky. For example, in

Massachusetts, an 1851 law established the public's right to access government documents. Those specific laws were refined over time, determining exactly what government documents the public has access to. In 1967, the Freedom of Information Act (FOIA) cemented the idea that the public should have access to government documents. What though, did that mean for personal documents issued by government agencies? Do those count as public record?

The answer is complex, and made more difficult by the division between federal and state level issuing agencies. Our personally identifying documents issued on a federal level, that falls to each individual state. Each state has its own view toward government transparency, personal security, and the public record. While the FOIA does not directly affect whether or not states disclose personal documents, many states model their regulations after the FOIA. This is dangerous because while the federal government does not have access to these particular documents, such as birth certificates and death certificates, the states do, and in turn, their citizens do.

This modeling comes into play in modern times with the advent of online ordering. The patterning after government openness bleeds into a security issue. In Massachusetts, "any person" meaning humans, corporations, and really any other requesting entity aside from animals, vegetables, and minerals, can request access to public records. While government transparency is typically a good thing, it is not such a great thing for state agencies to adopt that dogma when it comes to personal documents.

The birth certificate I ordered from Massachusetts is an excellent example of a state stuck in the middle between open and closed. When we began this process, we were operating under the assumption that open states gave unrestricted access to anyone's birth certificate who was born in that state, and that closed states did not allow anyone but family and court-appointed representatives to order birth certificates for individuals born in that state. This is untrue.

We searched for months to find the answer to a seemingly simple question: why are some states open and some states closed? The internet has some lackluster resources on this. The Secretary of State offices around the country which are technically responsible for the degree of how open or closed each state is, were equally elusive. As with so much of the research that went into this book, most of the information came from one lucrative phone call to the right person. In the case of open states versus closed states, the right person is Greg Sirko, Vice President of Sales and Marketing at VitalChek.

Whenever we have ordered a birth certificate or death certificate online, it's almost always through VitalChek. They provide the platform that government agencies use to order vital records. They make it really easy to figure out the information you need to place your order, and since it is the same company, it makes it easy for us to compare the process for ordering from state to state.

Mr. Sirko explained there is a vast spectrum from open states to closed states, which corresponded pretty well with what we saw in our own research. According to him, there are 56 vital record jurisdictions: the 50 states, Washington D.C., the United States territories, and New York City, which was granted its own jurisdiction. Within the 56 jurisdictions, there are over 6,400 individual issuing offices. Each of these issuing agencies is able to create its own version of a certified birth certificate, meaning that there are more than 9,000 potential formats for birth certificates.

One of our researchers, Chris Domingos, ordered two Massachusetts birth certificates for a family member. She ordered one directly from the town hall through an online portal, and one through VitalChek. Not only was Chris allowed to order the birth certificates and asked for no verification for herself (just as I was for my friend's birth certificate from Massachusetts), but the birth certificates looked completely different from each other. The certificate from the town hall is an obvious copy of the original, on nice paper signed by the City Clerk testifying that the birth certificate is real. The certificate from VitalChek is colored and looks like the form similar to the one I ordered through VitalChek for my friend. That one was signed by the Acting Registrar of Vital Records and Statistics for Massachusetts. Both versions are equally valid and can be used for the same purposes.

Aside from having two distinct-looking birth certificates, this particular case is even more complicated. Michael, the individual named on the certificate, was born as a surprise twin, and was called Baby A in the hospital. When Michael's mother later called Vital Records to name him, she wanted the middle name spelled Liem, but the registrar wrote it as Liam. She figured it out after they sent the birth certificate, and she later had to call again to request the spelling change. So Michael, has two, technically correct birth certificates, each spelled with a different middle name. That certificate, though not pictured here, presents an interesting question: in a sense, does Michael have two viable identities? Could he, or someone else open a credit card under those identities? Spelling differences are so often cast aside as typos, they are usually overlooked, and overlooking can mean everything.

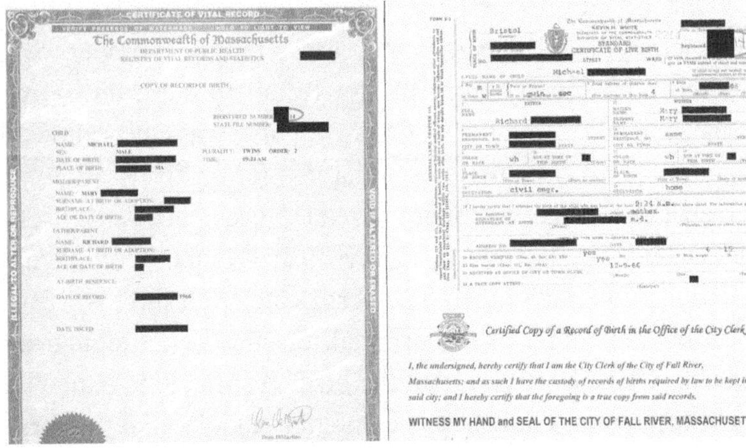

CERTIFICATE OF VITAL RECORD

The Commonwealth of Massachusetts

COPY OF RECORD OF BIRTH

Certified Copy of a Record of Birth in the Office of the City Clerk

I, the undersigned, hereby certify that I am the City Clerk of the City of Fall River, Massachusetts; and as such I have the custody of records of births required by law to be kept in said city; and I hereby certify that the foregoing is a true copy from said records.

WITNESS MY HAND and SEAL OF THE CITY OF FALL RIVER, MASSACHUSETTS

Michael's two different birth certificates, the one on the left is from VitalChek, the one on the right is from the town hall.

On the most basic level, we have two of the exact same birth certificates, for the same person, but they look totally different. There are several problems with this. First, how could any DMV employee or benefits program or individual who would use a birth certificate to establish an individual's identity know whether a document is real or not? "You would have to be really knowledgeable to tell the difference," said Mr. Sirko. According to him, birth records are kept in two places.

In Ohio, the state has no control over local offices. When a child is born, that child's information goes first to the local office where they register the birth, and then pass on that information to the state, who would in turn register it as well. According to Mr. Sirko, "Data entry in two places explains why the birth certificates look different. It is not unusual for somebody born in Cleveland to have a different birth date on their local and state issued birth certificates due to a data entry error, though as the registration process becomes more automated, they are avoiding this issue since both the local and state offices are actually working from the same database."

Would someone who has different data on each of their validated, government-issue birth certificates have two different identities, like with Michael's original birth certificate and the ones that Chris ordered? Mr. Sirko says yes. There could potentially be two different identities forged, and technically, neither would be fraudulent.

When we show people the birth and death certificates we have collected, people often assume they are fake. They squint at the seals and try to catch the reflective symbols embedded in the paper—without realizing that all these certificates are real. The only thing that is fraudulent is the relationship between the certificate and the person holding it. Why would anyone go to the trouble of making a fake birth certificate when it is so cheap and easy to obtain a certified copy of a real one? Upon seeing the piles of birth certificates we acquired, the CEO of a large data company asked how we got "all those fake birth certificates." He was shocked to learn they were real.

Creating a scale of the most open to the most closed states is hard enough as it is. When you take into account the additional territories and districts, it becomes mind boggling. Each issuing area has multiple (if not many) issuing offices, each of which vary in their approach. The small town agencies, such as those long-established town hall clerks, are generally more forgiving and will help you search for information rather than questioning your intentions.

As Chris found in both Vermont and Kentucky, the employees working at these agencies see it as their job to assist citizens in accessing that information. The mindset is similar to that of government benefit agencies: the mission being that they distribute benefits. Program success is generally determined by how much help they give, not how much help they withhold.

Chris encountered this helpful attitude in a phone call to the Kentucky Office of Vital Statistics. After a brief search, Chris had everything she needed to order a high-profile celebrity's birth certificate. We fully respect this individual's right to remain anonymous and will exclude her name from the narrative. Though doesn't it strike you as odd that we could be held liable for mentioning her name in a book, but having possession of her birth certificate is fine? We will call her Laura.

While the phone rang, Chris noticed that the website informs you of all the things you can do with a Kentucky birth certificate: "A certified Kentucky birth certificate can typically be used for travel, passport, proof of citizenship, social security [sic], driver's license, school registration, personal identification and other legal purposes."[2]

Chris was anticipating being shut down. Shouldn't the clerk tell her she has no business ordering Laura's birth certificate? Not only was Chris successful, but the clerk helped her obtain the information perhaps the internet leaves out. Here is an excerpt from Chris's email explaining the process:

> She asked if I was a relative. I said no.
> She asked what it was for. I said research.
> She asked for the father's name. I said I don't know.
> She looked it up for me.
> She said they never said no to anyone getting a certificate.
> She said they will do a search to help with the info.
> She said it is the law.

It reads like a poem, the imagined back and forth between a helpful clerk and an incredulous researcher. She said it's the law.

I'd never imagined wishing for a less helpful experience with a government office, but giving away personal information, to a stranger, for a celebrity? Part of why we order celebrities' birth certificates is to really test whether we can get them. Sure, if we order Joe Schmoe's, no one would know the difference. Ordering a personal document for a celebrity? It seems crazy. The helpful attitude of the people working in these offices can do some serious harm, especially if they believe it is more important to give out information than to safeguard it.

This attitude is changing with widespread fraudulent activity, and it is not a bad attitude for government agencies to have. When agencies strictly view their success in terms of how much funding, or in this case, information, they give out, they tend to leave out the caveat that their success should only be measured when that funding or information is given out properly and to the correct individuals.

It is heartbreaking to think of the many people whose lives have been severely affected because they were not able to receive badly needed funding, or whose personal information was carelessly given away. These agency-level errors look bureaucratic in the light of government reports or statistical findings, but for a student who lost out on funding or the newlyweds whose credit was unknowingly destroyed, this oversight can change everything.

The states that we have turned our focus on in this section are the ones that claim that they are open states. According to the site, an open state is a state where "anyone may order copies of [state] birth certificates, as long as they can provide the required information." These fully open states include

Kentucky, Washington, Ohio, and up until the day I was writing this section, March 30, 2017, Vermont.[†]

These open states allow anyone to order anyone else's documents as long as they can provide the information, such as the individual's full name, birthdate, parents' full names (including middle names and the mother's maiden name), and the place of birth. We tested this in many capacities, such as ordering a birth certificate with no middle name, the wrong birth town, missing or wrong parents' middle names, and still, it was sent. They were always sent. In fact, we have never been turned down.

The next tier of states we will call 'partly open.' These states, such as Massachusetts and New York, appear to have regulations in place, but those regulations seem to have no bearing on whether or not someone can order a certificate. In Massachusetts, I ordered my friend's document and claimed I was her sister. According to VitalChek, only the person listed on the certificate and immediate family can order. Due to the lack of verification, on who I was, I was able to order and receive one.

Larry ordered his own birth certificate from New York State. They requested verification in the form of his license, which he scanned and sent into New York's vital records office. While it is great that they asked for verification, the problem is that the license does not have to match the human who sent it in. A scanned photo ID will not be able to capture all the technological safeguards built into the ID. There is also the problem that bartenders have to face as well: there are over 50 different kinds of driver's licenses. That means you have to

---

[†] The Green Mountain state is voting on the bill H.111, which will begin to restrict access to birth certificates. We will discuss this further in a subsequent section.

either know what every variation of driver's license, including their individual security features looks like, or go through a laminated binder and check the ID against the photocopy you have. Or, if you are busy, you just don't check.

Hypothetical situation: you are walking down the street in any New York town and you find a lost wallet. Inside the wallet you find a driver's license and want to use that license to assist in stealing identities. A driver's license is very helpful. It has a full name, a birth date, an address, and a picture of whoever it belongs to. You sift through Google, Facebook, and if it makes you work for it, you consult an ancestry website. You might even have to go to the exhaustive trouble of subscribing to a white pages site. No one can verify if the person holding the license matches the picture on it. In this case, bars have tighter security since they have to actually compare your face to the face in the picture.

Finding parents' names is not hard. Finding out the town an individual was born in is not hard. Ordering their birth certificate is not hard. Now you have their information and a copy of their license. You scan that license, send it to New York's Department of Health (where they deal with vital records) and now you have a birth certificate. If you are the same gender and could pass for the age on the certificate, you now have the foundation of a new identity. The system for a partially open state like New York[†] requires more work than a fully open state, but not by much.

Some states are in the middle of the spectrum. In Georgia, no one but immediate family (or so they say) can order birth

[†] New York calls itself as a closed state. We classify it as partially open, as no states are fully closed in our minds. Calling a state 'closed' means we would not be able, in any form, be able to obtain a birth certificate we were not entitled to. We do not think that is the case in any state at the moment.

certificates, but anyone can order death certificates. Connecticut is the same. Some states have closed access for a certain number of years, like in Alabama, where birth records are closed for 125 years, but death certificates are only closed for 25 years. The list goes on and on, and with each variation comes different ways to work the system.

In places like Georgia and Connecticut, where death certificates are completely open, it comes with a whole other complication: Social Security numbers are included on the certificate. That means that the Social Security numbers of any deceased person are essentially public record. Many states have the same model that Arizona does for applicants ordering a vital record:

> Arizona Vital Records requires all applicants to submit a copy of their valid, government issued photo ID. You will be presented with an Authorization Form at the end of your order that includes specific instructions. When faxing your identification, please enlarge and lighten to ensure your fax is legible:
>
> - A state issued Driver's License that includes a photograph
> - A state issued picture identification
> - Valid Passport
> - Federal ID[3]

These requirements may seem stringent, but keep in mind, even if a state requires identification, it does not guarantee it is a legal transaction. Proof of identity does not equal identity, so why do our agencies act like it does?

Some states have fairly good security. When attempting to

order a friend's birth certificate from Illinois, I applied as if I was her, and they requested the corresponding Social Security number, which they would presumably run against a system like SSOLV, which is used by DMVs around the country to verify names and Social Security numbers. I have ordered several birth certificates at this point, and not a single time have I been asked for a Social Security number. This is great for Illinois, but it is a poor showing if asking for a Social Security number is the gold standard.

While we wanted to create a comprehensive database of state and other location-based policies regarding documents, we also understand that if someone's aim is to obtain a birth certificate, they probably would just order one from an open state. If they needed an identity in a closed state for some reason, they could just use the open state identity and "move" to the closed state. Not only would that make obtaining the primary identity easier, but it would better safeguard against being discovered due to the poor communication about identities between states.

To help understand why open states are even a thing, we have highlighted one state in particular that is in the process of becoming closed. Vermont proposed and passed legislation to tighten their access to vital records. Understanding the legislative side of this process helps undercut the burning question of who allowed this all to happen in the first place. The basic answer is that no one did. No one just decided to have these states allow anyone to order anyone else's birth certificate. They evolved to do so, just as identity theft evolved to be a crime that no one saw coming.

Vermont in Flux

Vermont serves as an interesting case study for us, and not just because it is the home of maple syrup, flannel, and out of staters using saying "quaint." Vermont is neither an open state nor a closed state. Vermont is in transition, and not just to sweater weather. In the spring of 2016, the Green Mountain State passed legislation that would limit its access to vital records. According to a report prepared in 2015 by Richard McCoy, the Public Health Statistics Chief for the state, "Vermont lacks even the most fundamental limits and protections, such as the ability to require name and purpose from the requesting party[4] [for birth certificates]."

Before the vote passed, Vermont statutes lacked the authority to deny any request for a birth or death certificate, or even ask the reason why it was being requested. In Vermont, these certificates have been considered public record, and "to require the requesting party to provide their name, reason for the copy or to show identification would be considered an impediment to obtaining a public record."[5] Meaning, any kind of security they might implement would be illegal.

To make matters worse, there was no record or way of tracking unusual requests. One Vermont clerk told us of a case where several thousand birth certificates were ordered at once, and while knowing full-well it had to be fraudulent, the clerks were required to send out the certificates.

A similar case happened in Ohio, when one individual made a request for 4,577 copies of birth certificates by one individual. Despite their suspicions, the clerks had to provide the copies under their statute. Mr. McCoy's report is quick to point out that Vermont is no different in this regard.[6]

There is another 2015 Ohio case in which a man utilized both the birth and death certificate of a young boy and adopted his identity for 15 years. Stories like this one crop up all the time, and because there is no system to track unusual patterns of request, Vital Records prove to be no help to law enforcement. In Vermont, there is no limit to how many copies of a Vermont birth or death certificate[7] a person can obtain.

Tracking these numbers was also not allowed by the statutes. This meant that agencies could not keep track of the individuals ordering these documents. Maintaining information about the requestors is not required, though it would not be too difficult to build that into a pre-existing online ordering system.

The 2015 report lists multiple ways in which in-house staff can take advantage of the lack of security. While vital record offices are the ones that print the certificate, lots of people help supply the information chain that feeds them. Hospital staff could create birth certificates, funeral home workers could use the Social Security number of a deceased person before the SSA could note it as inactive. There are no background checks conducted on any of the individuals who are working with these legal documents and accessing the personal information included on them, like hospital and funerary staff.[8] Just because you trust your staff does not mean they are not committing fraud. "Vermont is not immune, as observed by the number of embezzlement cases in recent years,"[9] the report cautions.

On a wider scale, the United States Office of the Inspector General reported[10] that between 85 percent and 90 percent of birth certificate fraud is the result of genuine birth certificates held by imposters—the most difficult fraud to detect.[11] It is far easier to find a fake birth certificate than a real birth certificate in the wrong person's hands. What is important to take from this statistic is that real birth certificates can be used fraudulently. That real birth certificates can create a new identity for someone with a criminal record, or someone who can justify stealing benefits from those who need it, and detecting a real birth certificate in the hands of a different person, a different identity, can be a difficult maze to navigate.

On May 22, 2017, the governor of Vermont signed the bill H. 111 into law. H. 111 began its journey almost twenty years ago, when Representative Dennis Devereux realized that any one person could order any Vermonter's birth certificate. Rep. Devereux is a town historian and politician, and for the last twenty years has worried that Vermont's open-state policies put Vermonters at risk.

"I knew years ago that anybody could ask for a copy of my birth certificate and have it, and use it to get a passport, or a bank account in my name," Rep. Devereux said to me over the phone. "I have been concerned for years about protecting people's identities. Now it is worse than it was twenty years ago."

Rep. Devereux first started work on a vital records bill to help secure Vermonters' birth certificates eight years ago, but one town clerk's dislike of the additions kept it from passing. According to Devereux, most people were in favor of adding protections to vital records, but other, less popular items were tacked on. It was those provisions that held the bill back, not the vital records aspects. The Department of Health became involved, as did the archivists. These complications eventually killed the bill. "Every group had different concerns," noted Rep. Devereux. "It was very long and complicated. As we took testimony from people with concerns, the bill just got longer and more complicated."

Several years later, Devereux made a point to return to the

issue. He suggested they remove the more problematic aspects of the bill to attempt to pass it only with the components of the birth and death certificates. "All I wanted was to protect access to people's' birth certificates," said Rep. Devereux. "Kentucky and Ohio still have wide open access. We hope this kind of legislation will catch on."

The bill will not be set into action until 2018 at the earliest, and according to Rep. Devereux, the new bill will make it difficult to get a certified copy of a birth certificate who is not entitled to receive it. Skeptical as I am about any government security actions, it seems that Vermont is headed in the right direction.

He also emphasizes that while he knows the bill is not perfect, it adds consequences. It provides for up to a $10,000 fine for "for knowingly making false statements about vital records or unknowingly owning, selling, or falsifying them." In Devereux's words, "it gives this bill some teeth." Vermont's Department of Health has a team of investigators as well, and suspicious ordering of vital records will be recorded in a database in an attempt to combat identity fraud, so in the case of the individual ordering thousands of birth certificates, they would be flagged.

Vermont's new act also directs the state registrar to operate a Vital Records Alert System to track and prevent fraud, to match birth and death records, and to prescribe the contents and form of vital record reports, certificates, and related applications and documents.

H.111 will require Vermont to operate a digital registration system for birth and death certificates, and will no longer allow town clerks to register vital record events, preventing openings for both accidental errors and fraud. H. 111 also limits the people who are eligible for a copy to family members and court-appointed representatives, officially making Vermont a closed state as opposed to an open one. Finally, the act requires that applicants present some form of identification, which is logged in a central database. This database, maintained by the State Registrar[12] would help combat fraudulent patterns, since the database could track orders. Much still needs to be decided on the implementation front, which is where most fraudulent interests lie. It is all in the details. Will I be able to order online? If I order online, how much security will I have to wade through? Will I still be able to order the birth certificate of a stranger, simply by providing the right information?

Rep. Devereux knows there will be unintended consequences, but seeks to simply get the bill through at least to make it harder for fraudsters looking to steal Vermonters' identities. The in-fighting and inclusion of other aspects into this bill kept it from becoming law for many years. How many identities were compromised in that time?

One of the other issues Vermont is facing is a fraud that frankly, I had never even heard of: marriage certificate fraud. It was one of those times you smack yourself on the forehead and realize you have been preaching something that you did not even understand. How manipulative fraud can be, how insidious. Birth certificates, death certificates, of course marriage certificates. Why not?

In 2011, one town clerk in Brattleboro, Vermont noticed an influx of couples seeking a marriage license that did not seem all that interested in marrying each other. The clerk's insights set an investigation into motion, and uncovered a matchmaking service for profit. A woman, we'll call her Anna, would charge fees as high as $12,000 to pair immigrants from Brazil with

Americans. In her home she had $117,000 in cash and 61 gold rings at the ready.[13]

In terms of the sham marriages, three important points emerged. The American citizens paid to make the scams possible rarely faced any sanctions. In Anna's case, not one of the 32 American spouses was prosecuted, while all of the immigrants faced deportation.

The second point is that Vermont seemed primed for this to happen. Vermont's laws for obtaining a marriage license are the least stringent in New England. Town clerks are not required to ask for identification, and seldom do. Massachusetts, and every other state bordering it, requires proof of identity. According to Richard McCoy from Vermont's Public Health Statistics Office, "The sad thing is that nothing has changed. There have been no changes to the statutes or business processes. That means what you will read in these articles could be happening right now elsewhere in Vermont… or happen again someday in the future."

The last important part of this story is that the town clerk notified federal authorities numerous times about the suspicious couples and their wedding planner. Despite that, it took three years for federal agents to investigate and stop Anna's operation. Of the 32 marriages for which Anna was prosecuted, 22 were legalized by the Brattleboro-issued licenses. Other marriage licenses were granted in Connecticut and Massachusetts. In each case, the immigrants paid Anna as much as $12,000. Typically, the American spouses received half the money,[14] despite receiving no punishment.

The Americans involved in the scheme received no retribution and roughly $6,000, while the immigrants were often deported, Vermont had no laws and continues to have no laws requiring identification for marriages, and despite the town clerk alerting authorities multiple times about the scheme, it took three years to bring it down.

In that same Vermont town, a woman created a fraudulent marriage certificate and used it to receive over $69,000 in government benefits from the man she "married," who happened to be a murder victim. She is accused of forging the license in May 2012, backdating it to October 2010, to before her fake husband was killed, and submitting it to an insurance company.

The fraudster took out a marriage license from the town in October 2010, but there is no record of the license having been filed. The town clerk of Brattleboro said she usually checks on marriage licenses monthly, but that the fraudulent license "must have been forgotten about and never checked on."[15] Those examples are from the same town. If that is just one town in Vermont, how much is happening state-wide? Nation-wide?

With most state-issued documents, there is no federal operating procedure, or set standards. Each state has its own standards, though they may not be the most thorough. There are innumerable variations of birth certificates, death certificates, and marriage licenses, with state-to-state variations on driver's licenses and non-driver IDs.

These inconsistencies make it more difficult to identify anyone state to state. Bartenders and government agency officials alike have a tough time telling the real from the fake, and that only includes licenses. Birth certificates, on the other hand, are produced in over 6,400 different issuing offices and include 9,000 formats, with few unifying principles among them, aside from basic birth information, which is why the same birth certificate ordered from two different places can look so different. States also allow their records to be accessed with varying levels of security. The United States is unique in this regard, since most other countries tend to issue relevant identity credentials on the national level,[16] meaning they have a uniform set of identification nation-wide, and one security standard.

2

Document Aids

The truth is, your identity has already been stolen.
—Frank Abagnale

Having an individual's birth certificate, which you use in order to adopt a fraudulent identity, is compounded with the strength of a Social Security number. Everyone knows you are supposed to protect your Social Security number, but there are ways of obtaining Social Security numbers that no one is even aware of. If birth certificates are the key to identity, then Social Security numbers are the lock on the door.

In the states with more 'open' policies, the way to obtain a birth certificate is by having certain pieces of information about a person: name, birth date, parents' names, and birth city. In my friend's case, I already knew all this information. It felt a little like cheating, since I was ordering the birth certificate of someone whose information I did not have to find.

Unfortunately though, but this type of 'familiar fraud' is fairly common. Familiar fraud is identity fraud carried out between two people who know each other: parents from their children, siblings from siblings, wife to husband, and even between friends. The Identity Theft Research Center defines it as when anyone close to you such as a family member or friend uses your identifying information or existing accounts without your knowledge or permission to create new accounts, make purchases, or commit other similar crimes. Think about it: those are people whose personal information you either already know or have quick access to. In 2014, there were 550,000 reports of identity theft perpetrated by someone the victim knew.[17] That number is confusing when you consider the 2014 Javelin Strategy and Research report, which stated that familiar fraud has already affected as many as 500,000 children in the United States and over 2 million senior citizens.

This is why you have to take fraud statistics with a seed of caution: they are often based on cases that have been reported. In addition to not even knowing about the fraud, in cases of familiar fraud, family members are often hesitant to report it in order to preserve family harmony. Many individuals will pay off their imposter's debt. Do not do this. As soon as you begin to pay off the debt, you are claiming it as your own, making it harder to expunge identity theft charges, or other future charges that you may not be able to pay off.

Remember the caveat that those reports only come from people who were aware of the fraud, and also reported it. Those two factors keep a lot of familiar fraud from coming to light. Think of the infants and deceased who have their identities stolen. They are not reporting anything. Then think of the spouses and children and parents who find out their family member stole their identity, and unsure of what to do, let it go unreported. Lots of people find themselves in this situation, and understandably do not want to report their loved ones. When it comes down from disowning a family member or separating from a spouse, many will choose to keep their lives from falling apart over their credit score.

If you are not interested in stealing your family member's identity, there are other options. The best place to find all of this information in a public forum brings us from life to death: the obituary. Not all obituaries are created equal, but take a look at this example from Kurt Cobain's obituary.

> Kurt Donald Cobain was born at Grays Harbor Hospital in Aberdeen, Washington, the son of waitress Wendy Elizabeth (née Fradenburg; born 1948) and automotive mechanic Donald Leland Cobain.

In one sentence, we have everything we need to order his birth certificate, which we did. Famous or not, a good chunk of people's obituaries are available in their entirety online. Obituaries are the first place to look if you need personal information to order a birth certificate from. When a loved one dies, people write obituaries to illustrate the details of that person's life, to honor them. Unfortunately, people do not realize how much that information exposes. Searching any state and 'obituaries' (preferably one of the more open states for ease of ordering) will give you all the information you need. It is astounding. It is callous. It is also incredibly effective.

Perhaps the most striking case study in this book comes at the expense of a young woman we will refer to as Heather, whom I met in this very fashion. While we ordered many birth certificates to illustrate different points and to understand the ordering process in each state, Heather's story weaves together the many different aspects that point out how convoluted identity can become. Namely, how identity fraud can jump from silo to silo, from identity theft to tax refund fraud, from benefits fraud to corporate fraud. One of the most important takeaways is that the impact of documents, their availability and importance, affects every industry, every agency, and every potential solution.

Heather's was the last birth certificate I ordered, but she will come up again and again as we follow her identity from dormant document to the potential of corporate identity fraud. I tried to order a birth certificate that I could pass for. The first time, it didn't work, but it shed some light on how the process works that is very important. The first set is for a person named Briana from Washington. Since the first did not work out, I ordered a second, which was for Heather from Kentucky. Both of these individuals are now deceased. Briana passed away almost twenty years ago. Heather, much more recently.

This section is not for the faint of heart. It was even hard to write it. I tried to rationalize why I was writing about this, to help people learn about identity fraud, and potentially help prevent it—but that thinking failed to make me feel better. I met both Briana and Heather online while sifting through obituaries. I wanted to find someone who I could potentially pass for to follow the rabbit hole. Could I compile everything I needed to get a driver's license with someone else's name and my picture? Fortunately, the number of young, dead women in open states is small, which made it harder for me to find a match. I warned you, it's grim.

The process we will follow for Briana illustrates how death certificates can aid in ordering a birth certificate. Heather's case shows how the two combined can work together to create an identity. Two of the ideal characteristics to look for when searching for birth certificates is that the individual is an infant, or dead, or both. In other words, an individual who is unlikely to report suspicious activity. If this person is very young, no one will be looking into their credit file or claims on their identity for years. If that person is deceased on top of it, there is even less cause to check on that identity, unless that person was elderly and died recently. In that case, the family is often reporting to Social Security or Medicaid, etc.

From Death Certificate to Birth Certificate

Briana was exactly what I was looking for. I found her on an ancestry website. I had some success in Washington state by that point, I figured that is where I should start when it came to ordering strangers' certificates. Starting at the beginning of the alphabet, I thought of the most generic names I could: Ashley, Allison, Amy and common last names like Smith and Thompson. I input a birth date range close to my own. After 20 minutes of searching through birth data, I clicked on the data set containing death information: grave records, public death records, obituaries. And finally, there she was: the perfect identity. Briana's entry was interesting, not least of all because, there were two different entries for her.

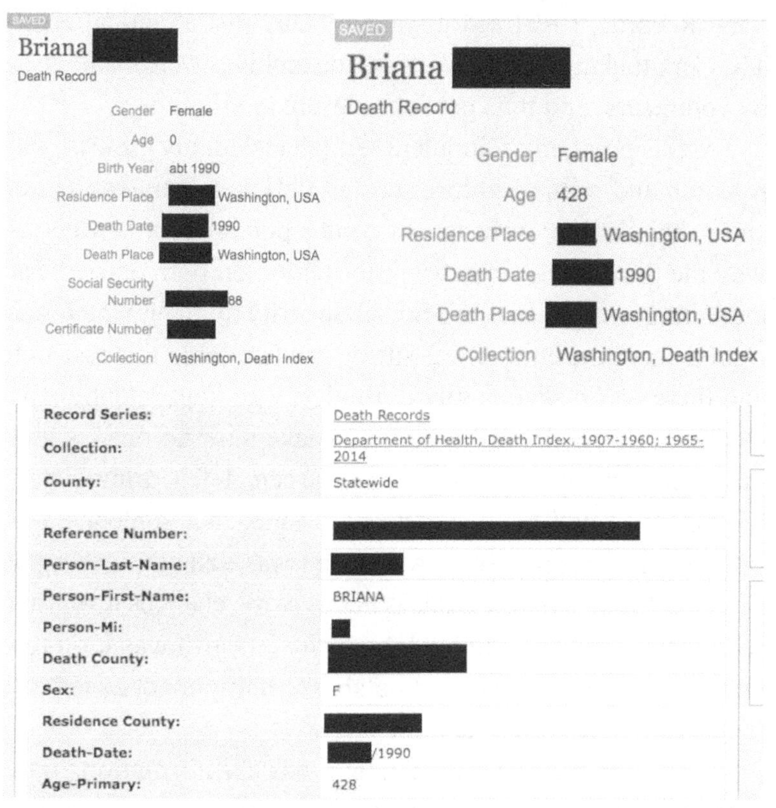
SAVED
Briana
Death Record

Gender	Female
Age	0
Birth Year	abt 1990
Residence Place	Washington, USA
Death Date	1990
Death Place	Washington, USA
Social Security Number	88
Certificate Number	
Collection	Washington, Death Index

SAVED
Briana
Death Record

Gender	Female
Age	428
Residence Place	Washington, USA
Death Date	1990
Death Place	Washington, USA
Collection	Washington, Death Index

Record Series:	Death Records
Collection:	Department of Health, Death Index, 1907-1960; 1965-2014
County:	Statewide
Reference Number:	
Person-Last-Name:	
Person-First-Name:	BRIANA
Person-Mi:	
Death County:	
Sex:	F
Residence County:	
Death-Date:	/1990
Age-Primary:	428

Briana's redacted entry on the ancestry website included her personal information that would eventually help me order her birth and death certificates.

The first two images come from the ancestry website. The third image comes from the Washington state digital archives, a public database accessible through the Secretary of State's website. Briana's name is the same on all three entries, though on the first, the town of residence is mentioned, whereas on the second it is only the county. The death date appears on both, but note the ages listed: on the first it lists Briana as being zero years old, and on the second, 428 years old. This struck me immediately, but after speaking with a clerk at Washington

Vital Records, I realized it was probably just a clerical error. The clerk told me that the Washington archives were catalogued by volunteers, and thus not very closely monitored.

At this point, my fraudulent self felt elated, my professional research and crime-fighting self felt deflated. It makes sense, since the ancestry website just combs public record. which is why the third image matches information on both entries. Far more interesting is Briana's Social Security number, which was right there in the entry. I was sitting in a bar doing this research and there was a skeezy guy to my left who kept intentionally elbowing me. I glanced around to make sure no one, skeezy guy included, was looking at my screen. I felt immediately guilty, like I had done something wrong, that someone was watching me. It's not like it was porn I was casually looking at in a bar, this was public data. What were the chances it was her real Social Security number? And why on earth was it on this publicly-accessible, state-run database? It might not have been illegal, but it definitely felt like it was.

I had an ideal candidate, her basic information, and, potentially, her Social Security number. This whole experiment began with the idea that finding enough of a stranger's personal information to order their birth certificate was possible, but it was starting to look like it was not only possible, but really easy. I could not find her parents' information, and I did not know her actual age when she died, but I did know enough to order a death certificate.

As the legislation for the state of Vermont points out, death certificates can be used to order birth certificates. Death certificates require less information to order than a birth certificate, and the death certificate contains far more information, including in many states, a Social Security number. Though birth certificates are the key to enabling proof of identity, death certificates contain enough personal information to make it worthwhile to order them.

Information Required to Order Certificates		
Birth Certificates	Death Certificates	
Full Name	Full name	
Birth date	Birth and death dates	
City and state of birth	City and state of death	
Parents' full names, including middle names and mother's maiden name	—	

After entering her name, death date, and location of death, I ordered Briana's death certificate. When it arrived, I tore open the envelope, and everything I knew about Briana was confirmed in an old-timey typewriter font. The Social Security number from the website was accurate, her death certificate confirmed it.

STATE OF WASHINGTON
DEPARTMENT OF HEALTH

Public Health - Seattle & King County Vital Statistics

CERTIFIED COPY OF DEATH CERTIFICATE

Date Issued: 1/20/2017

STATE OF WASHINGTON DEPARTMENT OF SOCIAL AND HEALTH SERVICES
VITAL RECORDS
CERTIFICATE OF DEATH

LOCAL FILE NUMBER

STATE FILE NUMBER

1 NAME—FIRST MIDDLE LAST: Briana
2 SEX: Female
3 DEATH DATE (Mo. Day Yr.): 1990
7 BIRTHDATE (Mo. Day Yr.): 1990
8 BIRTH STATE (if not in USA give country): Washington
9 CITIZEN OF WHAT COUNTRY: U.S.A.
13 SMOKING IN LAST 15 YEARS? (Yes/No): N/A
14 MARITAL STATUS: Never Married
15 SURVIVING SPOUSE: N/A
16 WAS DECEDENT EVER IN U.S. ARMED FORCES? (Yes/No): No
17 SOCIAL SECURITY NO: 88
18 HIGH SCHOOL GRADUATE? (Yes/No): No
19 USUAL OCCUPATION: N/A
20 KIND OF BUSINESS OR INDUSTRY: N/A
21 Was Decedent of Hispanic Origin or descent? 2 No
22 RACE: White
25 INSIDE CITY LIMITS? (Yes/No): Yes
33 BURIAL, CREMATION, REMOVAL, OTHER (Specify): Burial
34 DATE (Mo. Day Yr.): 1990
36 LOCATION—CITY/TOWN, STATE: Washington
37 FUNERAL DIRECTOR SIGNATURE: X

TO BE COMPLETED ONLY BY CERTIFYING PHYSICIAN

40 TO THE BEST OF MY KNOWLEDGE DEATH OCCURRED AT THE TIME, DATE, AND PLACE AND DUE TO THE CAUSE(S) STATED.

SIGNATURE AND TITLE: X Pathologist
42 DATE SIGNED (Mo. Day Yr.): 90
43 HOUR OF DEATH (24 Hrs.): ? AM

TO BE COMPLETED ONLY BY MEDICAL EXAMINER OR CORONER

41 ON THE BASIS OF EXAMINATION AND/OR INVESTIGATION, IN MY OPINION DEATH OCCURRED AT THE TIME, DATE, AND PLACE AND DUE TO THE CAUSE(S) STATED.

SIGNATURE AND TITLE: X

49 NAME AND ADDRESS OF CERTIFIER—PHYSICIAN, MEDICAL EXAMINER OR CORONER (Type or Print)

50 PART I. ENTER THE DISEASES, INJURIES, OR COMPLICATIONS WHICH CAUSED THE DEATH. DO NOT ENTER THE MODE OF DYING, SUCH AS CARDIAC OR RESPIRATORY ARREST, SHOCK OR HEART FAILURE. LIST ONLY ONE CAUSE ON EACH LINE.

IMMEDIATE CAUSE (Final disease or condition resulting in death). Sequentially list conditions, if any, leading to immediate cause. Enter UNDERLYING CAUSE (Disease or injury which initiated events resulting in death) LAST

(A) — INTERVAL BETWEEN ONSET AND DEATH: Immed.
DUE TO, OR AS A CONSEQUENCE OF (B)
DUE TO, OR AS A CONSEQUENCE OF (C)

51 OTHER SIGNIFICANT CONDITIONS—CONDITIONS CONTRIBUTING TO DEATH BUT NOT RESULTING IN THE UNDERLYING CAUSE GIVEN ABOVE
52 AUTOPSY? (Yes/No): Yes
53 WAS CASE REFERRED TO MEDICAL EXAMINER OR CORONER? (Yes/No): NJA 90-1273

61 REGISTRAR SIGNATURE: X
62 DATE RECEIVED: 1990

DSHS 9-150 (Rev. 1-89) -1187-

Briana's death certificate contained her Social Security number, the same one featured on the archive website.

Let me repeat that: Briana's correct Social Security number was on the death certificate, a certificate that anyone can order with the right information, which is available for free on a public archive or obituary. It didn't matter that I already had her Social

Security number. If I didn't already have it, I would still have gotten it with her death certificate, this number that is supposed to be a secret. How could it possibly be this easy? Again I had the feeling that someone was looking over my shoulder. It just felt wrong. Later, I realized that doing an image searching for 'death certificate' yields countless Social Security numbers, names, addresses, and anything else you could possibly want to perform identity fraud.

At first I felt shock, seeing Briana's Social Security number so publicly available, but then I started looking at the rest of the death certificate. My heart sank. In the fraud industry, people love to talk about how it seems like a 'victimless crime,' meaning there is no direct victim, how in a sea of identities it is easy to lose sight of the person whose life you are impacting. I've heard a lot of people say: "Someone who steals identities would probably never pull a gun on someone." I always believed that myself—had read it in enough government reports to internalize and regurgitate it.

Briana's death certificate changed my thinking in an instant. I barely noticed her parents' full names, the whole reason I ordered it in the first place. Briana lived for less than 30 days. Noted at the bottom in a barely legible scrawl were the words: Sudden Infant Death Syndrome. Small numbered boxes with obvious answers for an infant remind you of everything she never experienced.

Box 14 reminds you she never got the chance to meet someone wonderful, after a few terrible ones probably, fall in love and get married. She would never do that. She never got to figure out how she felt about the military, let alone serve. An N/A in Box 13 tell us she never got to be peer-pressured to smoke a cigarette behind the high school gym. Box 18, with a firm "No" tells us she never even went to high school. Even more sobering than the cause of death is Box 31, where the name of the informant of her death matches the name of her mother.

Briana's death certificate feels uncomfortable in my hands, not only because it lists her Social Security number, the lock on her identity, but because it feels like I have an intimate glance into a family's tragedy—a family I will never meet. Briana died 27 years ago, but these details paint a vivid picture. I see her tearful mother calling, in a near catatonic state, to report her death. I see the humble funeral, her father unresponsive to the rest of the family's attempts at consoling remarks.

After that, I can no longer subscribe to the notion that identity thieves can easily remove themselves from their victims. Holding the details of a tiny life in my hands, private intimate details that feel so clearly wrong, so clearly invasive for me to have in my possession. I do not believe you can disconnect yourself from these details. This crime is violent, but only to those who can feel it.

With Briana's death certificate in hand, I ordered her birth certificate. While these issuing agencies believe they are safeguarding their sensitive information, what they do not understand is the connection between the two. This is what makes states like Connecticut and Georgia stand out to me—places where the birth certificate is closed and the death certificate is open. Safeguarding the birth certificate will not do you much good if you can get the death certificate.

When Briana's birth certificate arrived, I felt a strange flurry of success mixed with the sadness of the situation, of feeling intimately familiar with the tragedy of strangers. The successful half plummeted immediately when I saw a giant 'DECEASED' printed on the bottom of the birth certificate.

What's more, Briana's name was even spelled differently. On her birth certificate, it has two N's.

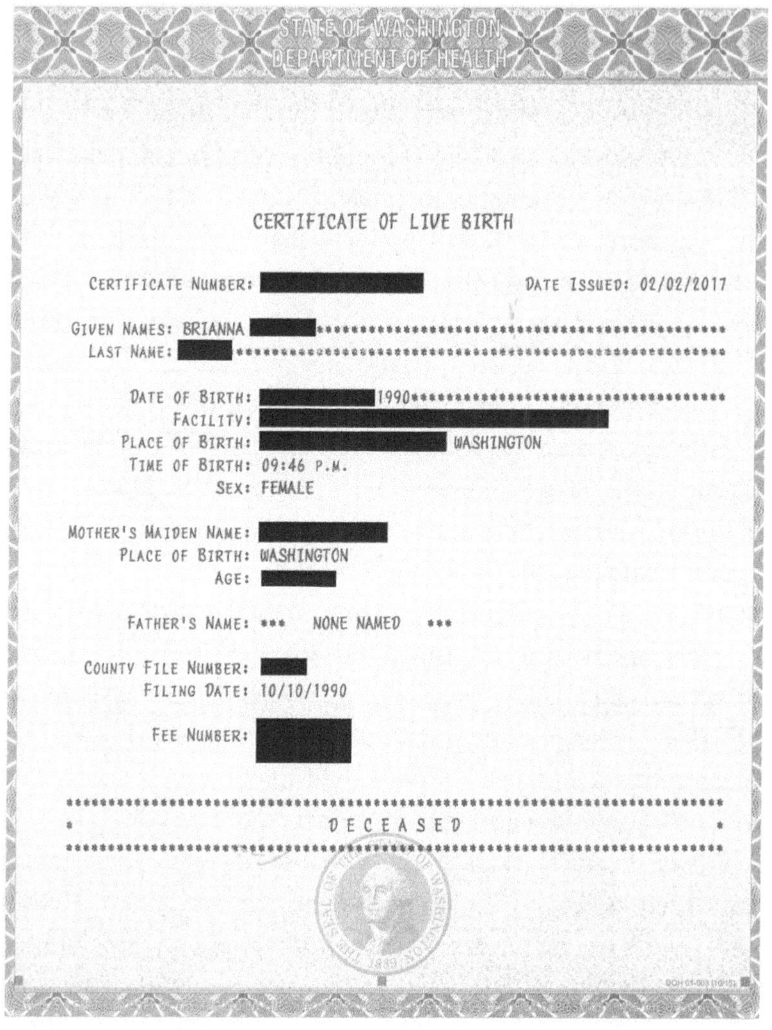

STATE OF WASHINGTON
DEPARTMENT OF HEALTH

CERTIFICATE OF LIVE BIRTH

CERTIFICATE NUMBER: [redacted] DATE ISSUED: 02/02/2017

GIVEN NAMES: BRIANNA [redacted] **
LAST NAME: [redacted] **

DATE OF BIRTH: [redacted] 1990**
FACILITY: [redacted]
PLACE OF BIRTH: [redacted] WASHINGTON
TIME OF BIRTH: 09:46 P.M.
SEX: FEMALE

MOTHER'S MAIDEN NAME: [redacted]
PLACE OF BIRTH: WASHINGTON
AGE: [redacted]

FATHER'S NAME: *** NONE NAMED ***

COUNTY FILE NUMBER: [redacted]
FILING DATE: 10/10/1990

FEE NUMBER: [redacted]

****
* D E C E A S E D *
****

THE SEAL OF THE STATE OF WASHINGTON 1889

Briana's birth certificate, with the notice that she is deceased.

This is one of the positive steps states are taking to help fight fraud. After an issuing agency receives news of a death, all birth certificates issued after that are marked as deceased. This is a great step—you cannot use a birth certificate that says 'DECEASED' and use it to get a driver's license, thanks to data-matching services. This also meant that the road I had thus been on, to attempt to gather all the documentation I would need to theoretically become Briana, had hit a dead end.

As we dove deeper into the rabbit hole of document fraud, we discovered yet another loophole. Briana died the same year she was born, in 1990, which meant that when I ordered it in 2017, the vital record offices in Washington had access to her death certificate for 27 years, plenty of time to archive her death certificate, locate her birth certificate, and mark it as deceased.

The time frame for matching birth and death certificates is far shorter than 27. Another birth certificate, Doreen from Washington, gave us a clue to how long it takes. Doreen died in mid-November 2016, and when I ordered her birth certificate in early January 2017, it was still unmarked. This tells us that a month and a half was not enough time to mark a birth certificate.

Any attempts to prevent fraud are good steps. This kind of system is called a 'front-end' approach. These systems do not play catch-up, as many others in this industry do. We will explore this idea later, but we would like to introduce it now to help showcase just how interconnected everything in this world is. Every part of fraud will interact with every other part. From here, you could jump to the chapter that talks about pay and chase, or the last chapter on implementing solutions. Front-end solutions, systems, and approaches will come up a lot. This may seem disorganized, but so is fraud. To commit it, there are a million elegant ways to do so. To try to categorize it feels impossible.

Fraud preys on complex government bureaucracy, on the

fact that we do not double check or verify. Fraud looks for the holes in systems and exploits them. Fraud thrives on chaos within systems, so when we talk about it, it feels unorganized. Fraud runs the gamut from document fraud, to large-scale hacks, to everything in between. There is old school fraud like check fraud, and new school like social media identity hacking. How can all of this fit under the single umbrella of 'fraud?' We wanted to mirror this, so when a different facet of fraud presents itself within the discussion, we will not shy away from it, but will eventually meander back to the original topic.

Even though systems are being implemented to mark birth certificates after an individual dies, it takes a good bit of time to match the certificates, leaving a window of vulnerability. This window is when people looking for birth certificates can troll obituaries, and order the birth certificate of someone who died that day, that week, that month, or even that year, and still feel confident it will be unmarked when it shows up in their mailbox.

Another point, in addition to how long it takes for a deceased individual's birth certificate to be marked as such, is that both Briana and Doreen's birth certificates are from Washington. Washington is only one of 56 issuing zones, that means there are 55 ones with completely different rules and policies. Additionally, that both Briana and Doreen were born and died in Washington, which means that Washington only had to communicate with itself.

Let's say Briana was born in Washington (which she was), but then died in Oregon. Is Oregon going to call over to Washington Vital Records to tell them Briana died? Will Washington then say thank you very much Oregon, you are such a helpful neighbor, and then mark Briana's birth certificate as deceased? Unlikely.

Most likely, it increase the time window in which you could order Briana's birth certificate before the matching system could mark her birth certificate as deceased. Would the birth certificate ever be marked? If it did get marked, it means those states can communicate with each other about vital records. Keep in mind when we mentioned Briana and Doreen were only born in one out of 56 jurisdictions. That means there are 1,540 unique relationships that have to be formed between districts, and yes I did have to relearn the formula for combinations to bring you that number.

Either that absurd number of relationships have to all be formed, or one centralized database with state-reported death data would have to exist that all states can report deaths into and extract death data from. That last sentence will be incredibly important for you to remember. Write down this page number now so you can check back later to prove you remembered it. The overarching idea that no one communicates well enough exists in every agency, in every business, and in every household when someone did not do their fair share of dishes and being passive-aggressive is your only coping mechanism.

The rabbit hole of documents, verification, and communication goes deep, with many tunnels branching off the sides. If states do not communicate with each other, it leaves loopholes for fraud. Hypothetically, if Briana died in Oregon instead of Washington, and I ordered her birth certificate, they may not have marked it as deceased. I would then be in the clear—with an unmarked birth certificate of a dead infant, who no one would probably check on, an accurate Social Security number, and a squeaky clean credit line.

Social Security Numbers and the Tie to Documents

Matching Personal Identifying Information (PII) including your full name, birthdate, and Social Security number, create the picture of identity to any agency that wants to know who you are. The SSA invented Social Security numbers to keep track of its beneficiaries. As a Secure ID Coalition report notes, (the bold and italic are theirs), "most of the credentials used to identify individuals are ***issued at the state level***." Most countries issue identity credentials on a national level, that and government benefits too. The brilliant report writers at the Coalition point out that the 50 different state/multiple territory approaches "are proving to be as problematic as building on sand."[18]

Since the United States has never been one for a national identity card, institutions and government agencies have a hard time keeping track of individuals, especially since a lot of people have the same name or birthday or move or change their name. People, and the identities that represent them, are not static. When the U.S. Tax Department, now called the IRS, realized there was essentially a national identifier for all working people, the Social Security number, it seemed like the perfect way to keep track of their people, as in taxpayers rather than SSA beneficiaries.

Usually, sharing data is a good thing. In this case, sharing Social Security numbers with the IRS was slightly catastrophic. The IRS began using the number as an identifier, despite the fact that the number has nothing that should qualify it to be used for identifying a person. The SSA protested, but the Tax Department piggybacked off their work and began assigning numbers to everyone. Not only did the SSA give them to working adults, but they enticed parents to get Social Security numbers for their children by tying it to a tax discount. And thus, tax fraud was born.

Millions of fake children appeared, and their real families were given a rebate, before the Tax Department required children to have a Social Security number in order to get the rebate. After that requirement, many of the children magically disappeared. This debacle turned the Social Security number from something that only working Americans had to something newborns had right from the start. It was not long before other institutions, government and private alike, joined in and started using Social Security numbers as a way of identifying everyone. Banks, schools, landlords, and credit card companies all got in on the action, and with their combined input, the Social Security number, though not at all actually secure, became a unique identifier for each person, in a sense, a national ID card. It also allowed institutions to check with each other, such as the banks against the credit bureaus, to confirm their customers.

Due in part to Americans' resistance to a national ID card, we ended up with something that is essentially the same as a national ID card, but without any of the security features built in to either the card or the number. Social Security cards used to have "NOT FOR IDENTIFICATION" printed on them, but they gave up after it became apparent that they were constantly being used for identification despite SSA's best efforts. Your Social Security card has fewer security features than your library card.

Much of this abridged history of the Social Security number comes from a video a friend sent me, which provides an excellently thorough and surprisingly funny account of the history of the SSA. The parting line gives a good sense of the video's tone:

> And that's the deal with this Social Security card: containing a national number for citizens that don't want one, on an identification card that fails at identification, given to all citizens except when it isn't, for a program that's universal, except when it's not.[19]

Since Social Security numbers have been used at length for identifying purposes, despite printing an explicit disclaimer on their older cards to not do that very thing. Using Social Security numbers for identifying purposes has opened up a raft of fraudulent opportunities. On its own, Social Security numbers can be used to steal identities, create fraudulent tax returns, assume a new identity, and hundreds of other schemes. There are most likely schemes happening that have yet to be discovered, and will start making headlines only after people start losing money.

A Social Security number adds to the complete picture of an individual identity, which gives a strong connection to document fraud. If you have a corresponding (or seemingly corresponding, as is the case with synthetic identities, which we'll get to) Social Security number and an identity-verifying document, such as a birth certificate, you have the foundation of a fraudulent identity. It is also far easier to pick an identity whose death certificate you can get a hold of, rather than a total stranger, and try to backtrack and somehow find their Social Security number.

People often ask where I find Social Security numbers, if not from death certificates. Common-sense has us thinking these documents are surely safeguarded, but this information is readily available. All you have to do is open your eyes. I used to be a bartender. At that job, a stack of employees' W-2s would sit next to a register for months, waiting for people to pick them up. Not *in* the register, but next to it, where anyone, employed or not, had access to it. They are helpfully branded as "Tax Information" to make it easier for anyone looking to get their hands on sensitive information.

How those W-2s could be used is an entire other chapter, but for the moment let's remember that they contain names, birth dates, Social Security numbers, and addresses, and that that information is very helpful when applying for a driver's license.

There are lots of other ways of finding Social Security numbers. Ancestry websites post them, as we saw with Briana's. Google images has thousands of pages of sensitive documents that list them. The National Archives has databases of veterans, with their Social Security number, birthdate, and death date listed. . Try it. Go to aad.archives.gov. Click 'Numerical Identification Files (NUMIDENT).' Pick a database to search, type in a generic first or last name, and you will have hundreds of PII at your fingertips in an instant.

From the victim's perspective, stealing identities is confusing. It's like looking out at a sea of data and trying to see the exact spot where the ship went down. It is seamless to perpetrate: with information more available than ever, it is hard for victims to know where their identity thief even found it. There is no beginning, middle, and end to the process. There are thousands of variations of how to do it, from death certificates to W-2s to forms left in an unlocked filing cabinet. Anywhere there is an opening, any document with personal information, should be regarded with just as much importance as your credit card.

Briana's death certificate listed every piece of information I needed to order her birth certificate and adopt her identity. This is the connection between documents and identity fraud. Since fraud presents itself more readily and makes for splashy clickbait when it is tied to giant data breaches, documents are left out of the conversation.

Documents are often left behind in the discussion about identity fraud, ignored for more "urgent" threats, like data breaches. I by no means intend to downplay data breaches, they are really bad. But by ignoring documents, by ignoring what now seems like an outdated and old-fashioned thought process, we are blinding ourselves to a big part of the problem.

From Birth Certificate to Death Certificate

One of the easiest and cheapest ways to gain access to seemingly safeguarded PII, is through death certificates. In Kentucky, I can get a name, Social Security number, birthdate, and address for $15, and it is perfectly legal. I found Heather's obituary from Kentucky with a single search. All sensitive information will be redacted from this book, but rest assured, all documents and obituaries featured are real. You can probably order them yourself. Here is Heather's heavily redacted one. Anything italicized was information that would help obtain her birth certificate:

> *Last name*, Heather *middle name*, *age*, mother of *children*, passed away *Friday*. Born in *city, KY* on *birthdate*, she was the daughter of *father* and

> *mother.* In addition to her daughters and parents, she is survived by *brother, grandmother,* as well as several aunts, uncles, nieces and nephews. She was preceded in death by *brother; paternal grandfather,* and *maternal grandparents.*

Heather's mother's maiden name was not included, however, the last line of the obituary lists Heather's maternal grandparents, which means we already know the mother's maiden name. With one search I had everything I needed except for the father's middle name. This was not as much as a problem as I expected it to be, since VitalChek advises providing anything but an *exact* match of what is on the certificate, may result in you not getting it. This is where death certificates come in. While Heather's obituary listed almost all the necessary parts, including her mother's maiden name through her maternal grandparents, neither of her parents' middle names were listed. Naive to the lack of stringency when it comes to ordering certificates in an open state, I ordered her death certificate in the hopes that it would list her parents' names in their entirety.

To order a death certificate in Kentucky, all you need to know is the individual's full name, birth date, date of death, and location of death. Even a basic obituary contains this information, so I ordered her certificate in the hopes that it would aid me in ordering her birth certificate. I knew from ordering Briana's death certificate that Heather's could potentially contain a wealth of information.

Heather died on a Friday, and I placed the order for her death certificate the following Tuesday, figuring since the obituary had been available online for a few days, surely her death certificate would have been filed. A week went by, my mailbox was empty. Two weeks, three. I finally called the Kentucky vital records offices. The clerk on the phone seemed surprised to hear from me. "Oh yes, we'll send you her certificate as soon as we get it."

"Sorry, what do you mean as soon as you get it?"

"Well ma'am," she said. "It typically takes at least 12 weeks from the time of death until our office actually receives the death certificate. She very well might be dead, but we won't technically know about it for a while. But we'll send you the certificate as soon as we have it!"

She was oddly chipper. Twelve weeks? The internet knew about Heather's death within 24 hours of it happening. After my experience with Briana, it dawned on me there was no risk of Heather's birth certificate being marked deceased. How could they mark her as being deceased if they did not know she was deceased, even if everyone else did? Even if random strangers on the internet did? I felt like I knew a secret I shouldn't—which is a part of identity fraud no one seems to talk about, just how creepy it feels. I imagine the people committing these crimes ignore this feeling, but just like with Briana, I felt that I had been privy to something deeply personal. If the state did not know about her death yet, why should I?

Pushing the feelings of invasiveness to the back of my mind, I hung up the phone and took a chance ordering Heather's birth certificate, since I still did not know her parents' middle names. Lo and behold, her unmarked birth certificate showed up a few days later, parents' middle names and all. Apparently, if you get most of the information correct, they will still send the certificate. Remember, their job as clerks is to help people get information, not withhold information. Finally, I had a birth certificate that matched my own description with an appropriate age range and gender. Who could argue that?

Registrar of Vital Statistics
Certified Copy

THE FACE OF THIS DOCUMENT HAS A COLORED BACKGROUND - NOT A WHITE BACKGROUND

FORM VS NO. 2-A (Rev. 9/87)

COMMONWEALTH OF KENTUCKY
DEPARTMENT FOR HEALTH SERVICES
REGISTRAR OF VITAL STATISTICS
CERTIFICATE OF LIVE BIRTH

FILE NO.
REGISTRAR'S NO.

Registration District No. 500 — Primary Registration District No. 2165

CHILD
1. CHILD'S NAME (First, Middle, Last): Heather
2. DATE OF BIRTH (Month, Day, Year): 1989
3. TIME OF BIRTH: 5:45 P M
4. SEX: Female
5. CITY, TOWN, OR LOCATION OF BIRTH:
6. COUNTY OF BIRTH: 234
7. PLACE OF BIRTH: X Hospital ☐ Freestanding Birthing Center ☐ Clinic/Doctor's Office ☐ Residence ☐ Other (Specify) — 1
8. FACILITY NAME (If not institution, give street and number): Central Baptist Hospital 08

MOTHER
9. MOTHER'S MAIDEN NAME:
10. SOCIAL SECURITY NUMBER:
11. AGE (At time of this Birth): 26
12. BIRTHPLACE (State or Foreign Country): Kentucky 415
13a. RESIDENCE-STATE: Kentucky
13b. COUNTY: 234 005
13c. CITY, TOWN OR LOCATION:
13d. STREET AND NUMBER:
13e. INSIDE CITY LIMITS? (Yes or No): Yes
13f. MOTHER'S MAILING ADDRESS (If same as residence, enter Zip Code only):

FATHER
14. FATHER'S NAME (First, Middle, Last):
15. SOCIAL SECURITY NUMBER:
16. AGE (At time of this Birth):
17. BIRTHPLACE (State or Foreign Country): Kentucky

INFORMANT
18. I certify that the personal information provided on this certificate is correct to the best of my knowledge and belief. Signature of Parent or Other Informant ►

CERTIFIER/ATTENDANT
19. I certify that this child was born alive at the place and time and on the date stated. Signature ► W. David Hoyer
20. DATE SIGNED (Month, Day, Year): 89
21. CERTIFIER'S NAME AND TITLE (Type/Print): Name W David Hoyer ☐ M.D. ☐ D.O. ☐ Hospital Admin. ✓ C.N.M. ☐ Other Midwife ☐ Other (Specify)
22. CERTIFIER'S MAILING ADDRESS (Street and Number or Rural Route Number, City or Town, State, Zip Code): KY

DEATH UNDER ONE YEAR OF AGE — Enter State File Number of death certificate for this child.

23. REGISTRAR'S SIGNATURE: Vivian O. Webb
24. DATE FILED BY REGISTRAR (Month, Day, Year):

THE BACK OF THIS DOCUMENT CONTAINS AN ARTIFICIAL WATERMARK - HOLD AT AN ANGLE TO VIEW

I, Paul F. Royce, Registrar of Vital Statistics, hereby certify this to be a true and correct copy of the certificate of birth, death, marriage or divorce of the person therein named, and that the original certificate is registered under the file number shown. In testimony thereof I have hereunto subscribed my name and caused the official seal of the Office of Vital Statistics to be affixed at Frankfort, Kentucky this 24 day of February, 20 17.

Paul F. Royce
State Registrar

Heather's birth certificate could serve as an identifying document for me if I wanted to use it.

Now that I had an indisputable certified copy of a birth certificate, my goal was to pull together the documentation necessary to apply for either a driver's license or a non-driver

ID card. I currently live in Wyoming, and recently went through the process of switching from a New Hampshire to a Wyoming driver's license, so I knew the process. In small towns like the one I live in, the DMV is friendly, local, and not nearly as anxiety-inducing as the ones in cities like Boston or New York.

I should note that when I exchanged my New Hampshire license for a Wyoming one, they used my Wyoming voter registration card as proof of residency. To get that, I had shown up in the office, shown them my New Hampshire driver's license, and pointed to my house out the window of the county clerk's office (it was actually right across the street) and explained that I wanted to vote in the primary. They gave me my registration without any further questions. Small communities are far from immune to identity fraud. Arguably, the lack of security in towns like mine make its population an easier target.

Three months after I ordered it, Heather's death certificate arrived. Just like Briana's, it contained her Social Security number.

Registrar of Vital Statistics
Certified Copy

THE FACE OF THIS DOCUMENT HAS A COLORED BACKGROUND - NOT A WHITE BACKGROUND

KENTUCKY CERTIFICATE OF DEATH

116

Case #:

To Be Completed By: Funeral Director (Must Be Typed)

HEATHER

N/A

FEMALE

2017

40

1989

KENTUCKY

N/A

CAKE DECORATOR

GROCERY STORE

02/09/2017

KERR BROTHERS FUNERAL HOME

To Be Completed By: Medical Certifier

2017

APPROX. 1130

CAUSE OF DEATH

DUE TO (OR AS A CONSEQUENCE OF):

DUE TO (OR AS A CONSEQUENCE OF):

DUE TO (OR AS A CONSEQUENCE OF):

3+ MINUTE(S)

Natural | Accident | Homicide | Pending Investigation | Suicide | Could not be Determined

UNKNOWN

UNKNOWN

ADMINISTRATION OF DRUG LISTED

03/16/2017

801

CORONER

C. DAVID (SONNY) RANSDELL

MERCER COUNTY CORONER, 348 EAST LEXINGTON STREET, HARRODSBURG, KY 40330

Paul F. Royce

03/16/2017

THE BACK OF THIS DOCUMENT CONTAINS AN ARTIFICIAL WATERMARK - HOLD AT AN ANGLE TO VIEW

I, Paul F. Royce, Registrar of Vital Statistics, hereby certify this to be a true and correct copy of the certificate of birth, death, marriage or divorce of the person therein named, and that the original certificate is registered under the file number shown. In testimony thereof I have hereunto subscribed my name and caused the official seal of the Office of Vital Statistics to be affixed at Frankfort, Kentucky this 20th day of March, 2017

FORM VS NO. 1-A
(REVISED 06/2015)

Paul F. Royce
State Registrar

Heather's death certificate, like Briana's, contained her Social Security number.

It was so easy to get these pieces of information, it felt a little surreal, especially since I had already acquired the most difficult components to obtain. I wanted to compile everything I would need to apply for a driver's license under an assumed identity, an identity that no one would check on, no one would question, and I was already halfway there.

To apply for a driver's license in Wyoming the requirements are as follows:

(1) Proof of Identity
- Birth certificate
- Passport
- Driver's License or non-driver ID card

(1) Proof of Social Security number
- Bank statement
- Social Security card
- W-2

(2) Proof of Wyoming (or other state) residency
- Utility bill
- Rental lease
- Mail addressed to you at your house

I had my proof of identity in the form of Heather's birth certificate, plus her Social Security number. There are lots of ways that I could have established my Social Security card. I could have gone to my bank, filled out a form (using my new birth certificate), and created an account, which would then allow me to get bank statements. I checked my own bank, and yes, you can apply for an account online.

Instead of potentially committing a felony, I took a more simple route. I downloaded a W-2 form from the IRS website, filled in my old job's Employer Identification Number (EIN) and financial information, my address, and Heather's name and Social Security number, and then hit print.

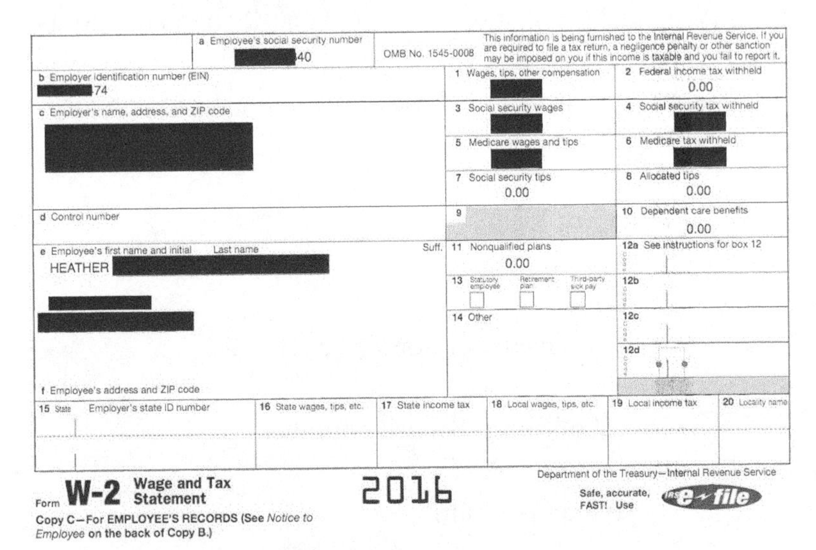
a Employee's social security number
40
OMB No. 1545-0008
This information is being furnished to the Internal Revenue Service. If you are required to file a tax return, a negligence penalty or other sanction may be imposed on you if this income is taxable and you fail to report it.
b Employer identification number (EIN)
74
1 Wages, tips, other compensation
2 Federal income tax withheld
0.00
c Employer's name, address, and ZIP code
3 Social security wages
4 Social security tax withheld
5 Medicare wages and tips
6 Medicare tax withheld
7 Social security tips
0.00
8 Allocated tips
0.00
d Control number
9
10 Dependent care benefits
0.00
e Employee's first name and initial Last name Suff.
HEATHER
11 Nonqualified plans
0.00
12a See instructions for box 12
13 Statutory employee Retirement plan Third-party sick pay
12b
14 Other
12c
12d
f Employee's address and ZIP code
15 State Employer's state ID number
16 State wages, tips, etc.
17 State income tax
18 Local wages, tips, etc.
19 Local income tax
20 Locality name
Form **W-2** Wage and Tax Statement
2016
Department of the Treasury—Internal Revenue Service
Safe, accurate, FAST! Use IRS e-file
Copy C—For EMPLOYEE'S RECORDS (See *Notice to Employee* on the back of Copy B.)

Making a W-2 for Heather was incredibly easy.

All that was left was proof of residency. The Wyoming Department of Transportation's website states that "any piece of mail showing your address will suffice." That seemed questionable, so I tried to stick with more formal modes. I had a WiFi bill that I scanned and edited so it was addressed with Heather's name and my address. The second piece of proof of residency I used was my own copy of my rental lease. Since it was my copy, and hadn't signed it, I wrote in Heather's name. Proof of residency, check. I also could have printed out any lease template and filled that in, but I liked the authenticity.

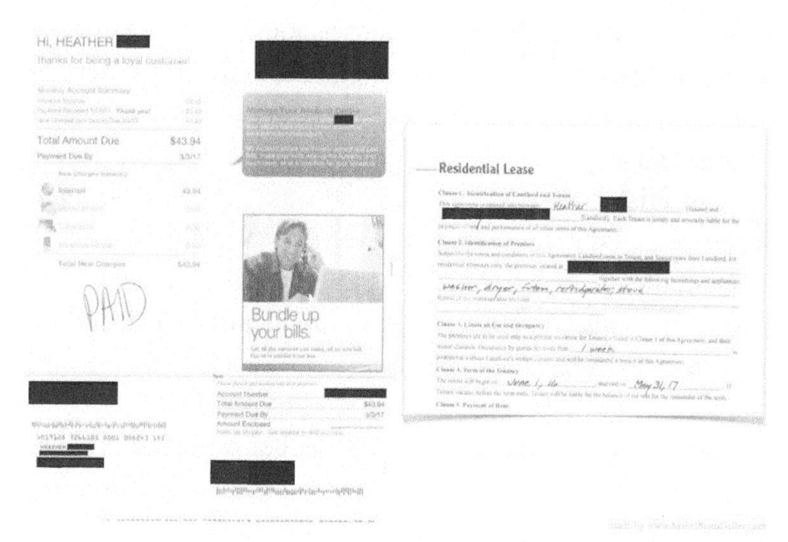

These altered identity-verifying documents would prop up Heather's 'identity' and allow me to obtain a driver's license in her name.

I compiled all these documents in under 45 minutes without a fancy scanner, printer, or special paper. At a conference in the winter of 2016, I showed these documents to a special agent in the FBI. She had served with the agency for over 20 years, and when I showed her my documents, she was appalled. She put her head in her hands when I explained that while I had to forge my WiFi bill, the birth and death certificates were real.

These requirements—proof of identity, proof of a Social Security number,[†] and two proofs of residency—are common for most states when it comes to applying for driver's licenses. In fact, they are the requirements mandated by states to order REAL ID compliant licenses. The REAL ID act was first proposed in 2005 as a way to streamline the necessary

[†] It should be noted that not every United States citizen has a Social Security number, since not all citizens are required to pay into the Social Security program. The exceptions usually correspond with old rules about unions and specific religious groups.

requirements to apply for a license, and how those licenses would appear. Now it forces some states' citizens to get new licenses if they want to continue using that form of identification to board an airplane.

Essentially, the REAL ID act was supposed to make American identification bulletproof. The problem is that I can meet all the requirements needed to get a REAL ID compliant license without actually matching my identifying document, making the Department of Homeland Security's claim that "every state has a more secure driver's license today than before the passing of the Act,"[20] questionable. For some reason, birth certificates are still acceptable forms of identification, even though they have nothing on them to tie the person to the document. In some states, you can order any birth certificate with ease.

So say I did it. Say I took my birth certificate, my W-2, my lease, and my bill, and applied for a driver's license. Hypothetically, there are systems in place to flag my application. These systems are the lynchpin to my hypothetical, fraudulent enterprise. If these systems work correctly, I would not be able to establish a fraudulent identity, nor use that identity to go forth and swindle every agency and institution in my wake. Establishing my identity, with the documents I have compiled, enables me to go on to test these systems. So what are these systems? Strangely enough, the process begins when someone dies.

3

Death Data and the Verification Dilemma

Bunny Lebowski: Uli doesn't care
about anything. He's a Nihilist.
The Dude: Ah, that must be exhausting.
—*The Big Lebowski, 1998*

The Social Security Administration (SSA) is responsible for keeping a record of everyone who has a Social Security number. Keeping track of these numbers seems like a simple: (A) Person exists. (B) Person has Social Security number. (C) Person and their Social Security number are stored in a database. This, like much in government, is just not how things pan out. The way the SSA has historically kept names and Social Security numbers organized is the 'Numident,' a file in which a big long list of everyone's name and number corresponds, and there is just one record for each person, even if you have had a name change. The SSA's mission, according to their website, is "to administer national Social Security programs as prescribed by legislation in an equitable, effective, efficient, and caring manner."

Nowhere in that mission statement does it mention

categorizing and logging death information, because frankly, that is not what the SSA's job is. The SSA is a benefits agency. They give out Social Security benefits, and the associated number, as we discussed, was simply a tool used to help keep track of those benefits. Its convenience was too hard to ignore.

The Social Security number has been used and abused by many agencies for many other purposes, including for identity purposes (which, as we noted, was a terrible choice). The Social Security number was never intended to be used for identity and keeping track of individuals, it was meant to be used as a way of keeping track of benefits.

It was convenient for other agencies to begin using the Social Security number for other purposes, in the same way that the Numident is now used to keep track of everyone's death data—even though that was never its intended purpose. You'll notice a trend where agencies try really hard to share information, but when they do, they completely misuse it, and get irrationally upset at the results. If you borrowed your neighbor's cake pan to bake muffins, and then got mad at your neighbor because you did not end up with muffins, "well that's just like, your opinion man."

It is equally ridiculous to be angry at how the SSA's information, their metaphorical cake pan, does not fit other agencies' criteria. If the SSA has a cake pan, and they only make cake, why would they own a muffin tin in the first place? The better question is why don't their neighboring agencies understand that a cake pan won't make muffins, and go buy their own muffin tin? This analogy is complicated by the fact that the SSA has a greater financial incentive to collect death information for beneficiaries than for non-beneficiaries.[21] This is so logical it's painful, but papers reporting on the pitfalls of the SSA's death data point it out like it's heretical. It's some serious victim blaming.

The SSA is more focused on the death information of beneficiaries because again, they are a benefits agency. If their mission is to distribute benefits, they obviously would keep closer track of those individuals. There is no reason for them to even keep track of data for non-beneficiaries, aside from having miserly, muffin-loving, non-data-keeping neighbors.

The only reason the SSA has death data of any kind is to verify if its recipients are still alive. It is no good sending benefits to dead people, though it happens all the time. In fact, a 60 Minutes episode called "Dead or Alive" focused on the consequences of the errors within the SSA's death data. In it, the producers note that,

> it's deadly serious business because when you're added to the file, that means that banks, the IRS, Medicare, law enforcement and the like, scratch you out of existence…And then, there are those who are on the Death Master File who are very surprised to hear that they're dead.[22]

It is quite literally the Book of Death, because once you are inscribed in it, there is no going back. One of the women discussed in that episode explained that she has to provide a letter from the SSA (which is updated monthly) every time she goes to the bank, just to prove she is alive. Every institution you interact with is safe to assume you are attempting to steal a deceased person's identity, not that you were accidentally put into a database of dead people. It seems that anything the SSA holds sacred, both the Social Security number system and the Numident, quickly become appropriated and exploited.

With that noted, we can move on to the three tiers of government death data and just how everyone else misuses it. This is not to say that the SSA is innocent, far from it, but just to pepper in the fact that this mess is not entirely their fault. Essentially, the SSA comes up with a great system for their own organization, other agencies see those systems and think, "Oh wow, what a great system I think I'll take that," to which the SSA responds, "Woah Nellie, that system shouldn't be used like that," to which the agencies react by totally ignoring them. These other agencies abuse the SSA's technologies, so exactly what technologies are they?

Hogtied by the Social Security Act

What database an agency uses tells us what information they have access to, which heavily affects the payout of benefits and ability to commit identity fraud. These databases vary in both their completeness, and to what degree the information within the databases is verified. The CSI Effect has contributed to the overwhelming misunderstanding of how verification works within government. The belief that all databases are created equal, that they are up to date and contain a complete set of information, and that verification exists where it does not, is false.

You would think if someone dies, their information will be recorded on a federal list of deceased people. Not necessarily. This is sometimes due to bad practices, but more often it comes down to a lack of funding, manpower, or available information. Having an up-to-date database, and verifying all of that information takes time and money.

By looking at identity fraud through the law enforcement

lens, if someone is stealing $1,000 a year, the effort it would take to bring that person in would end up costing more than just letting them steal it, making stopping that crime a waste of money. This is one of the main reasons why the frauds that make headlines cite astronomical amounts of stolen money: because it was worth the time to track down, arrest, and prosecute. This begs a few questions: just how much low-level fraud is occurring? When you put all the low-level fraud together what kind of a financial loss does it create? How much time and emotional strain does it collectively put on people and businesses who have had their identities stolen?

The havoc identity fraud can wreak is substantial, and one of the basic reasons why understanding death databases is important. Death data is one of the tools we can use to fight against identity fraud, by cross-referencing names and Social Security numbers with deceased individuals whose identities may have been compromised. Deceased individuals are ideal targets for fraud, and if our way of accounting for them is limited, so is our ability to use that data to fight fraud.

Venturing into the murky depths of the SSA's various death databases is no small task. Understanding the SSA, its utilization of the Social Security number, and finally other agencies' historical abuse of their hard work, is all background information on the journey to death database enlightenment.

The SSA was created during the Great Depression as a way of ensuring that citizens would have some sort of funding for their later years in life. The Social Security number was merely the way they kept track of what people put in and took out of the Social Security fund. Remember, the Social Security number and card have no identifying factors such as a picture, biometrics, or even a description of the person it belongs to.

The lack of identifying features is what makes it a poor way to tell who is who.

When the SSA first assigned numbers, cards were issued in offices nationwide. In 1972, the SSA began exclusively assigning cards through a computer-based system from its headquarters in Baltimore, Maryland, changing over from a paper system to an electronic one that could be accessed anywhere.

This switch from individualized, local offices to one centralized, electronic database created the 'Numerical Identification File,' also known as the 'Numident,' which is the master file of Social Security number holders. The Numident contains all the same information those local offices once held, just electronically and in one place. It, hypothetically, lists one entry per identity, and contains that identity's name, Social Security number, any history of name changes or card issuances, and if they are deceased.[23]

Note the use of "identity" here rather than "person." This is because, to an electronic file, there is no way to tell whether an identity is actually tied to a person or not. Your identity is how government agencies, banks, credit bureaus and other institutions know who you are. To them, that data is a more legitimate picture of identity than appearing in person. Therein lies the issue—when dealing with a seemingly matching pair, how do you differentiate a person from an identity?

The first potential for error was when the local SSA offices transferred their pre-1972 Social Security numbers to Baltimore to be added to the electronic Numident. Think of the childhood game Telephone: usually, the word you start with gets transmogrified into something completely different after being whispered between six different people. Data works the same way. The more times data is transmitted, especially by hand, the more room there is for error, both in translation and typed. Expecting to never have errors in a database is unrealistic, but we can expect decent verification checks to be sure that clerical errors are caught and fixed.

After a death is reported, the SSA runs that identity's information against the Numident file. If a matching file is found, the SSA then marks the Numident with a death indicator.[24] This death information was not publicly available until 1980 when, in response to Freedom of Information Act requests, the SSA began to extract Numident records with a death indicator into a separate file called the Full Death Master File[25] to keep all the deceased identities separate and for easy access.

When someone dies, their death information arrives at the SSA in a variety of ways. Their death data comes from funeral homes, relatives, other federal agencies, and state vital record offices.[26] The importance of who reports death information and who they report it to cannot be understated. This is the crux of death databases, because depending on those two factors, information may not actually trickle down to the other death data files. This is because of something called the Social Security Act, which has a certain provision that essentially ties the SSA's hands:

> This is because the Social Security Act requires the SSA to share death information, including data reported by the states, with federal agencies to ensure proper payment of benefits to individuals. This same act also prohibits the SSA from sharing state-reported death information for any other purposes.[27]

This creates a strange little loophole: on one hand, the SSA must share death data with federal agencies so they can cross-check their beneficiaries against the list of potentially deceased people. If the death data reported to the SSA comes from a state, however, it is not allowed to be shared at all. This is problematic for a couple of reasons, but most obvious is if state-reported data is missing from the database agencies are using to see if their beneficiaries are actually alive or not, they are not getting accurate results.

Consolation Databases

As a result of this security measure—allowing agencies to check for dead, and potentially fraudulent, beneficiaries and still protecting states' death data—the SSA maintains two versions of the Death Master File, the 'Full Death Master File' and the 'Partial Death Master File.' They are like the phases of the moon, except instead of tides depending on them it's identity security and the integrity of benefit programs, and that instead of cycling back around to being full, each of these files gets progressively more deficient the further along you get.

These two terms are of my own invention, as the SSA rarely differentiates between them except in dense reports. Often, they are lumped together with no distinction between the two. Most often, news articles and networks call the Partial Death Master File the 'Death Master File,' or the DMF when they are talking about the file that most agencies are able to access. For clarity, we'll use the two terms and discuss the differences between the two.

The Full Death Master File contains all death records seen on the Numident. This makes it sound like the Numident and the Death Master File are the same thing, but they aren't. The Numident is just a list of all Social Security recipients, which keeps track of deceased beneficiaries. Only the deceased beneficiaries are transferred onto the Full Death Master File.

The Full Death Master File would be a great resource, if agencies actually had access to it. As of December 2015, only nine agencies were receiving the Full Death Master File. One of the main requirements for eligibility is that the agencies pay federal benefits. This requirement bars some important agencies from having access to decent death data. The nine agencies with access to the Full Death Master File include:

- Centers for Medicare & Medicaid Services
- Department of Defense
- Department of Veterans Affairs
- Internal Revenue Service
- Office of Personnel Management
- Department of Agriculture
- Federal Retirement Thrift Investment Board
- Pension Benefit Guaranty Corporation
- Railroad Retirement Board[28]

This list of Full Death Master File recipients is more interesting when you realize what agencies are not on it. The Department of Homeland Security? Nope. The United States Justice Department? No way. That's because they are not benefits paying agencies, and thus not worthy of having decent death data. Those agencies get to use the Full Death Master File's inferior younger sibling, the Partial Death Master File.

Following the Social Security Act, the SSA created the Partial Death Master File, which is exactly the same as the Full Death Master File, but without the state-reported death records. This was a bad idea because without those deaths, the Partial Death Master File has 10 percent fewer records than the Full Death Master File (roughly 87 million, compared to 98 million).[29] By having fewer deaths included, it inherently becomes a worse file. The other reason is that aside from those nine agencies that get to access the good file, every single other agency or institution uses the bad one.

Those agencies include everything from government agencies like the Department of Homeland Security to state-run DMVs, to private businesses like hospitals, universities, and credit bureaus. So if someone tried to take out a credit card in Michael Jackson's name, and the credit bureau ran the stolen Social Security number through the Partial Death Master File, they would know that (A) Michael Jackson is dead, (B) that the person attempting to order a credit card was probably not Michael Jackson, and was likely attempting credit card fraud, and (C) maybe ghosts exist.

Over 700 agencies and institutions have access to the Partial Death Master File. While some make a lot of sense, like Experian, one of the three major credit bureaus, others not so much. Even though the Partial Death Master File has fewer deaths, it still has roughly 87 million deaths, which is a significant number when it may not be in the best hands. Why does the Painters Union Pension Fund need to have access to 90 percent of dead Americans' Social Security numbers and other personal information? Employers have the capability to run their employees (which is highly recommended) through a database to be sure they are using their actual identity, but they can do so without having access to the Partial Death Master File. The SSA's website allows you to upload names and Social Security numbers to verify them for wage-reporting purposes.

So again, why does Cooling and Winter, LLC, Attorneys at Law need access?

It's okay though, because these groups pay for the shoddy data. Groups can subscribe to the Partial Death Master File through the National Technical Information Service. NTIS reimburses the SSA for the cost of providing the file, and then sells it through a subscription service. It isn't a one-time purchase, well, to put it bluntly, because people keep dying. And since people are dying every day, with over 2 million dying every year, it is unrealistic to expect such a vast and ever-changing collection like American death data to ever be totally complete and up to date. Agencies that need access to the full file can't get it, agencies that do not need it can access the partial file because they pay for it, and agencies that should have access to the full file but don't have to pay for access to the bad file. It's all gone higgledy-piggledy. The easiest way to understand how these databases relate to each other, and how this in turn affects anything, is to follow the process of officiating a death. Naturally, for research's sake, that is exactly what we did.

Through the Death Data Maze

The key to this entire exercise is understanding that every agency and entity is separate, and unless there is evidence of data sharing or verification, there may not be any. To help make these death data files more concrete, we will illustrate the events that occurred following the death of Evelyn, Larry's mother. We start with the arrival of her death certificate, and follow the trail to the eventual ceasing of her benefits. Tracing Evelyn's passing allows us to see the trickle-down of information through the

SSA databases, and learn what information is where and who has it.

When Larry ordered her birth certificate, Evelyn had already been dead for three years. Despite this, her birth certificate was not marked as deceased. Since it was not, it makes it clear that the vital records agency did not understand Evelyn was deceased. That understanding is very different from the SSA's knowledge, meaning a state-level agency is not privy to the same data that some federal agencies are.

Larry also ordered her death certificate from the state of New Hampshire as her son. New York City knows she was born. New Hampshire knows she died. Does New York City know she died? No, they do not. How would they begin to relate a birth certificate to a death certificate? The birth certificate has no Social Security number on it, and there could have been multiple people born in New York City with that same name. It starts to sound like a word problem, but the bottom line is this: if you order a birth certificate of a deceased individual who died in a state where they were not born, the certificate will not be marked as deceased.

Larry's mother passed away in 2013. She was born in Brooklyn, New York, lived in New Jersey for most of her life, and died in New Hampshire. The guidelines for ordering birth certificates in New York City are as follows, as they appear on VitalChek:

> You may order copies of New York City birth certificates for yourself or for your child, as long as your name appears on the certificate. Your name must also appear on the credit or debit card that is being used for payment. If your name does not appear on the New York City

> birth certificates being ordered, you must order
> in person or by mail.

The security caveat at the bottom would be an ideal step, but unfortunately ordering by mail provides the same shield as ordering over a computer. When Larry typed in Evelyn's name, he saw that part of New York City's verification process includes Knowledge Based Authentication (KBA) questions.

You are probably familiar with these types of questions if you use online banking, cloud storage, or if you have accessed your email on a different computer than usual. Knowledge Based Authentication questions prompt you with questions like, "Who was your first grade teacher?" or "What street did you grow up on?" These types of questions are an excellent first line of defense against fraud. The junior fraudster sees these and is immediately deterred.

Unfortunately, KBA does have its holes, and Larry found them immediately. KBA questions work well if someone attempts to steal the identity of a person they do not know, but if it is a 'familiar fraud,' and you know a lot of personal details about that individual, it isn't too hard to crack. Also, many KBA questions are combed from public record data. This means if the questions can be generated, a crafty fraudster could gain access to the same information.

In the vein of a familiar fraud, Larry was able to answer the system's KBA questions with ease. While he ordered her death certificate as her son, he posed as Evelyn herself to order her birth certificate. The system asked, "What area code do you currently live in?" They did not mean where did Larry currently live, they meant Evelyn. This question is not only easy to answer for someone who knew Evelyn, but also tells us that despite having her name, state, city/borough, and birth date, the system does not trigger a notification within VitalChek that the individual is deceased.

This turns into a technologic discussion. It does not appear that the vital record agencies that subscribe to the Partial Death Master File have the ability to review identities in real time. It is also evident that VitalChek does not receive death information in any filtered way. Again, this is understandable because VitalChek is not a vital records agency, but simply a way to access them.†

Cross-correlation between death and birth certificates, especially across state lines, does not happen. With common names especially, it becomes difficult to connect the death certificate to the birth certificate because there is no common way to link them. 'Boundary blindness' masks an identity's existence in another place and time, letting some live on well past their due.

When Evelyn passed away, the funeral home informed Larry that they would arrange for the death certificate, which would in theory alert the state of New Hampshire and the SSA that she had died. Assuming since the SSA had been alerted, and that all the correct agencies would in turn learn of her death through the Partial Death Master File, Larry was alarmed by the continued arrival of pension checks from New Jersey.

Larry called the Division of Pensions and Benefits for the state of New Jersey on three separate occasions and told them to stop sending the checks. New Jersey said that to stop the checks Larry had to cease Evelyn's Medicare benefits. This made no

---

† Remember, VitalChek is just the operating platform that places the orders. The requirements that appear on VitalChek are set by each state's security and privacy laws. Ordering a Washington birth certificate through VitalChek involves little to no security, whereas New York City's requirements are more stringent.

sense. Evelyn was already deceased, and was thus not receiving Medicare. Wouldn't Hospice, the government program that confirmed her death, have already informed them?

In the meantime, the SSA stopped her Social Security benefits immediately, raising the question that once they have her death data, why isn't Medicare able to access it? This reinforces that neither Medicare nor the Division of Pensions and Benefits in any state have access to the Full Death Master File, since the database they accessed did not contain Evelyn's state-reported data. If Larry had not called, the pension checks simply would have kept coming.

During these phone calls, Larry asserted that he was able to view Accurint, a file informed by the Partial Death Master File, and that she was still listed as alive. Larry called the SSA to double-check they had her death on file. The SSA asserted that they knew about it, which is why they had stopped her Social Security benefits. After the third call to the Division of Pensions and Benefits, they finally requested the money be returned.

The SSA has a policy that states once a beneficiary turns 100, they need to actually see the beneficiary in person. An agent is sent to make a physical check on the individual. Since Evelyn was 84 when she passed, New Jersey would have continued to send her checks for another 16 years, totaling about $300,000 that Larry could have potentially pocketed, and disappeared with.

We can trace the knowledge of Evelyn's death from agency to agency. New Hampshire Vital Statistics first told the SSA about her death. The SSA recorded this information on the Numident with a death indicator. Because the record then had a death indicator, it was inscribed on the Full Death Master File.

However, it was not inscribed on the Partial Death Master File because it was reported by New Hampshire Vital Records—a state—meaning that New Jersey Pension could not have received notice of her death. Since New Jersey Pension is a state benefits giving agency rather than a federal agency, they do not have access to the Full Death Master File, which also means that when they verify names and Social Security numbers, they are checking an incomplete list of information.

While the SSA's death data may not be the best, there are a couple of aspects that make it even worse: completeness, timeliness, and verification. Tracking Evelyn's death data exposes three major holes within our system. The completeness of the SSA's death data, how long it takes for death data to be recorded and utilized, and whether or not the death data the SSA has is actually verified all contribute to why exactly the SSA's death data may not be the best.

It's Not Me, It's You

The United States Government Accountability Office (GAO), whose job is self-explanatory, started noticing the holes in the Partial Death Master File in 2001, but they really got serious with a full-scale report in 2013. In that report they outlined issues the Partial Death Master File had with regard to both completeness and verification. In another report they published in October 2016, much of the language was exactly the same. Much of the research we used comes from 2013 GAO reports. Let that date range, 2001-2016, be an illustrative point: we can still use data from 2013, and even 2001, because it still applies, because despite those reports, not much has changed since then. In 2016 they literally quote themselves, which is a bad sign when it comes to radical changes made to an archaic and poorly functioning system:

> In November 2013, we reported that the SSA's processes for collecting and maintaining death reports could result in untimely or erroneous death information, such as including living individuals or not including deceased individuals. For example, we identified about 500 instances in which death reports submitted to the SSA in early 2013 listed dates of death in 2011 or earlier. This is of concern because—if the dates of death are accurate—the SSA and other agencies may have been at risk of paying benefits to these individuals for long periods after they died.

The GAO notes that the SSA has taken steps to improve their death data's accuracy, and that since 2014 they have completed a risk assessment and a data quality assessment, and have been developing a business process for the third phase. These plans return to the idea that, while it was not the SSA's fault in the first place, the burden has been placed on their shoulders, and since it may be a while until it is removed, they might as well try to make their data better.

With the excusable notice that much of the data we are using is old, but not outdated, that we acknowledge it and do not believe it matters since no legislative changes have been made, and finally that none of this should be blamed on the SSA in the first place since all they wanted was cake and not muffins, we will move forward. The three major areas of weakness within the SSA's death data include incomplete data, slow transmission, and fallible verifications. These issues contribute to why these databases are not accurate and trustworthy sources of information.

Incomplete Data

The GAO has been publishing papers and studies for years citing the same information: the Partial Death Master File has errors and a gap of information that is preventing agencies and institutions from properly screening their constituents. In these reports, there is also a severe lack of differentiation between the Full Death Master File and the Partial Death Master File. It is difficult to know whether that is due to a lack of knowledge that the Partial and Full Death Master Files are two separate lists, or simply a lack of specification. Either way, we have tried to parse the information to the best of our ability and talk about them in three distinct arenas.

Issues with the completeness of the Full Death Master File have been around for a long time. In a 2001 Social Security bulletin, it was noted that the Death Master File contained 17.5 percent fewer deaths than were reported in the United States Vital Statistics system from 1970-1991. This could be for a couple of reasons. The electronic Numident was not created until the 1970's, with paper records being transmitted to an electronic file in a fashion that most understandably could have been erroneous. Additionally, the Numident files with death indicators were transferred to the Death Master File in the 1980's.

Every time we transmit data, there is room for error. Every time we extract data, there is room for error. This is one of the biggest pitfalls of data collection. Rather than using a single repository for data, and restricting access to that and requiring better security, new files are getting created. Not only does this allow more room for transmission errors, but those files are less likely to be complete.

The older files are not deleted as new, ever-more specific ones come along. The SSA still uses the Numident for their internal purposes. Nine agencies use the Full Death Master File, and everyone else gets the Partial Death Master File.

There is not a lot of logic to why older files are not deleted, but my best guess is because the newer files are not actually better. Each data set serves a specific purpose, and blocks access to certain things because of certain legislation. For example, states' death data can't be included on the Full Death Master File, so instead, they created a whole new database. This model is inefficient, and by continually creating newer and worse data sets, the SSA is setting up an entirely new data file that has the potential to be abused. The more data that exists, the more data there is to steal.

It seems odd that, since the Partial Death Master File is so restricted in order to protect state security, that the death data is available to such a wide array of private institutions. What appears to be lacking in all this is an understanding that not just federal agencies need access to death data, while some organizations are not granted any access. Often, agencies and institutions request and keep far more data than they need. This creates the challenge of protecting and safely securing all that data, and if they do dispose of it, doing so in a safe and secure manner. Agencies and institutions call for more and more data, and when shared, it is shared with the wrong people. This process could be streamlined by agencies limiting their requests for information to what is relevant and necessary for their purposes.

In May 2013, the GAO reported an example of an agency that wasn't receiving the Full Death Master File, despite having good reason for access. The Department of Treasury would not be eligible for the Full Death Master File, which they needed to administer the Do Not Pay Initiative. In 2009, President Barack Obama signed an Executive Order with the goal of "Reducing Improper Payments and Eliminating Waste in Federal Programs." One of the ways they thought of to reduce improper payments was the creation of the "Do Not Pay List." This is a list of databases that include death data agencies should consult before they pay benefits, to make sure they are not paying benefits to the deceased. The Do Not Pay List includes the Debt Check Database, the Credit Alert Verification Reporting System, and yes, the Partial Death Master File.

A 2015 GAO report stated that the Do Not Pay Business Center, a component of the initiative, wanted access to the Full Death File. The request was denied, with no documentation aside from the brief explanation that, once again, the SSA cannot release state-reported death data.[30] Given all we know, ensuring agencies run their beneficiaries through death data is a really great idea. This is an excellent front-end approach, except for the fact that agencies are relegated to checking inferior databases.

In 2001, a report stated that 1.3 million deaths recorded in the SSA's Master Beneficiary Record (the master payment file for the Old-Age, Survivors, and Disability Insurance program) had not been entered into the Numident file, and so were not included in either versions of the Death Master File.[31]

In 2016, the Office of the Inspector General conducted an audit that discovered millions of deceased number holder's whose death information was not on the Numident. As a result, these number holders' information did not appear on the Partial Death Master File. Not only that, but from February to June of 2016, the Office of the Inspector General identified 8,689,084 number holders whose Numident record contained a date of death, but whose death data did not appear on the Partial Death Master File.[32] From 2001 with 1.3 million missing records on the Partial Death Master File to 2016 with about 8.7 million missing records, we can conclude that the problem of incomplete records is not only being perpetuated, but is increasing.

The way the SSA keeps these records might be partly to blame. There are certain data-keeping and recording practices begging for errors. The May 2013 GAO report states that if the SSA receives a death report and cannot match it to a Numident record because of differences in name, date of birth, or gender, it generally will not take actions to resolve the non-match. Since these entries are now problematic, they would not be recorded on the Partial Death Master File. That means that a single data entry error can keep a name off the list all our agencies use to check against. Other examples of erroneous entries include:

- 130 instances where the death date was earlier than the birth date
- 1,295 instances where a death occurred between the ages of 111 and 129
- 1,791 individuals whose deaths occurred before 1936, the year
  Social Security numbers were first issued. All of those individuals had Social Security numbers assigned to them.[33]

These are just a few examples of typos leading to death data not being included. How many typos exist that have not been discovered and remedied? There are 8.7 million death records on the Numident that are not included on the Partial Death Master File. Different databases corroborate these errors. They cite the following information to prove those number holders were deceased:

- Individuals who are 65+ despite having no earnings in the past 25 years or receiving payments from the SSA
- Nearly all 8.7 million Numident records contained entry code "D," indicating the number holders were already deceased when the SSA input their information onto the electronic Numident file in the 1970s.
- Approximately 5.3 million number holders also had dates of death on the SSA's Master Earnings File (MEF) and/or the SSA's
- The SSA received reports that individuals using 69,863 Social Security numbers had approximately $3.9 billion in wages, tips, and self-employment income between 2009-2014. Remember, these are Social Security numbers tied to identities that have already been proven to be deceased.

The Numident, the Full Death Master File, and the Partial Death Master File are increasingly deficient. Since the latter databases rely on the former, at the very least they have to be worse. With 10 percent fewer records than the full Death Master File, the agencies accessing this version of the file, such as Homeland Security, the Justice Department, may be missing deceased program participants.

Agencies can request the Full Death Master File, and the SSA determines what agencies are eligible on a case-by-case basis. But, the SSA officials said they were not aware of any standards for how to determine who should be able to access the Full Death Master File.[34] So while there are standards and a way to become eligible to access the better file, no one knows the process for doing it.

One thing becomes clear over and over in these government reports: using the Partial Death Master File data to verify identities is not as surefire a tactic as people seem to think it is. The GAO notes that, "As a result, any benefit-paying agency relying on the partial Death Master File to help identify deceased program participants may be missing death records"[35] and that, "such inaccuracies could adversely affect the Death Master File's usefulness in helping agencies combat improper payments."[36] There are sections from these reports where they quote older versions saying, "as we reported back in 2011…"

Seeing how little has changed is frankly, embarrassing. Since 2001, the government has known there are problems with using this kind of death data, and we have revisited it several times since, so why is there still a problem? Aside from the slow-moving beast that is government bureaucracy, there is an even bigger problem. The organization that is getting constantly blamed has no ability to actually fix the problem.

We know there are problems in all three death archives, the Numident, the Death Master File, and the Partial Death Master File, but whose responsibility is it to fix these problems? The Office of the Inspector General states, "As a result, we believe that the SSA should revisit its decision to exclude these records from the Death Master File."[37] This single statement rolls all responsibility onto the shoulders of the SSA. This is why it is important to discuss Social Security numbers within this context: because the SSA should not bear the sole burden of other agencies misusing their inter-agency systems.

The SSA cannot, in fact, revisit their decision to exclude those records, since it was not their decision in the first place. They do not have the power to reverse it. In response to a GAO report about the Do Not Pay Initiative and using death data to help secure it, Frank Cristaudo, Executive Counselor to the Commissioner of the SSA responded in a letter explaining the SSA's situation. As you read through the next section, feel free to replace "DNP" with any government agency who should probably have access to reliable death data:

> Under current law, we are not authorized to share state death information with the DNP portal. Section 205(r) of the Social Security Act (Act) limits the purposes for which we may disclose state death information. While the Federal Improper Payments Coordination Act of 2015… requires us to share our death information with DNP, it did not amend section 205(r) of the Act to include authority for us to provide DNP State death information. Thus, we appreciate GAO's recommendation that "Congress should consider amending the Act to explicitly allow the SSA to share its full death file with Treasury for use through the DNP working system." This recommendation also aligns with a proposal in the President's fiscal year (FY) 2017 Budget that would grant us the legal authority to share all our death information, including data from the states, with DNP.[38]

While it is exciting to know that the President had a plan to expand the allowance of death data for 2017, it is arguably doubtful that any such expedient and logical moves will be made under the current administration. The letter from Social Security also acknowledged that "the best possible way to improve the quality of the death information we collect and share is for Congress to fund the continued expansion by the states of Electronic Death Registration."[†]

We can break this letter down into a few comprehensive parts. Mr. Cristaudo gently reminds the GAO that the SSA is not actually allowed to share state death information, and the reason for that is because of the Social Security Act. Mr. Cristaudo also mentions his appreciation for the GAO telling Congress to amend the act, since they still cannot do anything unless it has been amended. The most exciting and meaningful part of the letter comes at the end, where Mr. Cristaudo mentions that there is a proposal in the 2017 fiscal plan to grant legal authority so that the SSA can share state reported death data.

This letter from Mr. Cristaudo ties together all the things we have discussed in this chapter: that the SSA's hands are tied due to the Social Security Act, and everyone keeps getting angry at them for not including death data that they are not legally allowed to include. It is the Social Security number debacle all over again. What really should happen is that agencies stop stealing and using materials the SSA made for their own purposes, and expecting these materials to fit their needs.

A lot of these errors, such as checking to see if someone's death date comes after their birth date, are simply common sense. If someone were reviewing this information, they would be able to catch these errors simply by looking at the data. These types of errors, compounded with the inability to include

[†] We can whole-heartedly voice our agreement for this. The Electronic Death Registration System is a great front-end solution that could drastically cut down on fraud and abuse. We will discuss EDRS in full at the end of chapter four.

state-reported death data, are responsible for huge gaps in the system.

It is a simple equation: if the Numident is not complete, and the Full Death Master File is even less complete, and the Partial Death Master File is even less complete, agencies or businesses that subscribe to the database are, of course, not able to reliably access complete information. This means that if a subscriber, say a state's Department of Motor Vehicles were to actually run a name and a Social Security number, it may not even show up.

Slow Transmission

Let that sink in: the identity verification databases that agencies and institutions rely on are missing 8.7 million identities. Probably more. The second tier of problems with death data is timeliness. How quickly (or slowly) death data reaches vital records agencies allows us to learn things about how this whole machine works.

Remember Heather? After I ordered her death certificate, and waited in vain for several weeks for its arrival, I learned that the Kentucky Office of Vital Statistics did not know about her death. This meant that if I ordered her birth certificate, it most certainly would not be marked as deceased. That also tells me that the information trickle-down has probably been too slow for her death data to have reached the Partial Death Master File.

Why would that be important? Remember, the Partial Death Master File is what every other document-handling government agency or business uses to verify identities. If her death had not yet been recorded, it gives me, a potential fraudster, a certain window of time to establish myself within her identity: get a non-marked birth certificate, get a driver's license, order a credit card. That window of time might just be one of the most crucial elements of this multi-faceted equation.

Just how big that window is depends on a couple of factors. It certainly depends on whether the individual was born and died in the same state. If that is the case, odds are you would have to move faster to get established within their identity. As with Evelyn's death, someone whose lifetime spanned two different states, the Social Security Act serves as a buffer. Larry could have continued receiving her pension payments for another fifteen years before anyone came to check on her.

As we have noted, the SSA receives death notifications from a variety of sources, such as states and state-level agencies such as New Hampshire's Division of Vital Records, family members, funeral directors, post offices, financial institutions, and other agencies. These reports are usually received within 30 days of the date of death.[39]

A report from 2001 detailed the fact that death data usually takes about 90-120 days to actually reach the SSA.[40] According to that same report, death data reaches financial institutions within one to two months, depending on when the SSA receives it. According to these numbers, which may have changed since 2001, if Bob dies on March 15th, and his death is not reported for thirty days, so roughly April 15th, add in the 90-120 days for it to reach the SSA, we are in mid-August. Add another month or two until the financial institutions have it, which means we have from March 15th to mid-September (being conservative) before the financial institutions actually have his death data—and that is only if his death was reported by a non-state entity, because otherwise, his death data will not be included at all. If I wanted to apply for a credit card in Bob's name, I would have a safe window from March 15th, the day he died, to roughly September 15th, to do it in without being flagged.

How slowly death data moves within federal agencies is not a unique problem. As with Heather's birth certificate and death certificate, there were several months before the Vital Statistics office would have known about her death, even though I knew about it from an online obituary within 24 hours. Becoming familiar with these time tables allows fraudsters to know exactly what they are getting themselves into. There are time periods and financial ceilings in fraud to stay within. If you do stay within them, your odds of getting caught drop significantly.

If a fraudster is trolling obituaries, acting relatively quickly (within one to two months) will ensure that that individual's death data has not reached the SSA, or vital statistics offices within the state, and their information will not yet be recorded. While the funeral home or hospital dispatches the death data to state Vital Records and the SSA within a week or so, recording that information can take a while. The other way to ensure this, especially in the case of ordering birth and death certificates, is to find an individual who was born and died in two different states. Evelyn's death data highlighted to us exactly what agency had and did not have her information, and were we looking to fraudulently abuse her identity, it would tell us exactly what agencies to take advantage of and what agencies to avoid. The SSA was out of the question, since they already had her death data, but because of the Social Security Act, New Jersey's Division of Pensions and Benefits did not know about her death and continued to send her checks. All we had to do was nothing.

Providing more timely death information and disseminating it in a more thorough fashion would certainly not hurt, but it is not the entirety of this problem. In addition to a lack of complete information and the slow transmission of data, fallible verifications prove to be a serious issue.

Fallible Verifications

Verification with regard to document fraud is just as problematic as it is elsewhere. If it is difficult to verify the information an agency is presented with, it becomes even more difficult when the information the agency is using to verify said information is wrong or outdated. This is the case with the SSA's handling of American death data.

It feels surprising to hear this, to realize how little verification occurs, how little states communicate. Unfortunately though, this 'boundary blindness' exists in almost every sector of state-level governance. Boundary blindness means that where one state's boundary ends, so too does the jurisdiction of that state, and their ability to track information stops.

Boundary blindness is one of many issues states face when attempting to get good information. The types of security measures that protect these databases from false information, like a person who is alive being listed as deceased, can be limited to typo-finders. Even when security measures are in place, they are often significantly flawed and under-utilized. A common complaint from state workers is that they do not understand how to use new technologies put in place that help verify data.[41]

Within this already lacking framework lies the verification trap. A lack of verification works in both directions: agencies fail to verify, but so too do the people and businesses that work with them. There is an existing system where businesses have the ability to verify the Social Security numbers of new hires. A 2015 Office of the Inspector General report points out that less than 1 percent of the six million employers used this system in 2014.[42]

You'll recall the CSI Effect we have discussed a few times already. The idea has a home here, as well. If I were to bring Heather's birth certificate and the other documents I compiled to a DMV, they would run her name and Social Security number through the Partial Death Master File, verifying that Heather is dead and by association, revealing that I, the person attempting to get a license, am attempting to commit fraud. Let's say her death was not reported by a state, in that it would be allowed to be on the Partial Death Master File. Let's also say that her death data was transmitted quickly, skipping the approximate six month window before it would be recorded on the Partial Death Master File, and that I had waited a full year after her death to get a license.

Basically, say everything went against my favor, and that luck was on the side of the good guys. Say, that against all odds, everything lined up perfectly so that if all the information was where it should be, the doe-eyed DMV clerk would sit up straight at their desk and demand the strapping young security guard throw me in jail for attempted identity theft. Even if everything was in place, the system could still fail us, entirely because that doe-eyed clerk did not actually verify anything. Or, it could be that someone in the SSA did not verify the report and one number or letter was off on Heather's death data, which means that the system would not have included her in the first place, would then not recognize her, and nothing would be flagged. The lack of checks and balances is the verification dilemma.

While it may seem like we are beating a dead horse, it is important to remember that while the SSA's data is not perfect, it was never meant to be. Recall the section on Social Security numbers and how the SSA got blamed when every other agency mooched their seemingly-convenient data and got upset when it was not sufficient for their needs. The SSA created the Social Security number and the Numident to keep track of their beneficiaries to be sure they were not paying out benefits to the deceased. That being their sole objective with that data, there was no reason for them to keep track of non-beneficiaries, until the Tax Department was giving out Social Security numbers to help alleviate the mess of keeping track of individuals and taxes.

This truth makes the fact that the SSA does not verify non-beneficiary deaths a little more digestible, though still not acceptable. The changes that the SSA has been called on to make are necessary for this data to be reliable, but it does feel a bit like telling them to Be the Bigger Agency, like, "I know your little sister the IRS stole your toy, but understand that they are younger and you need to let them play with it even though they might break it and cry that it's your fault and several billion dollars of benefits are at stake." It is apparent, as well, that the SSA is sick and tired of being blamed for their faulty data. In addition to Frank Cristaudo's not-so-subtle letter, NTIS's website (the institute that distributes the Partial Death Master File) gets pretty sassy:

> The social security office is correct at this time in saying that the problem now lies with you (the subscriber of the Death Master File). In the latter case (2 above), the Death Master File subscriber (you) probably received the incorrect death data sometime prior to the correction on the SSA's main records. (The only way you can

> now get an updated Death Master File with the correction would be to again purchase the entire Death Master File file and keep it current with all of the Monthly or Weekly Updates—See Mandatory Requirement)

If you cannot hear the deep sarcasm in the bolded "See Mandatory Requirement" of the last line, you miss the alluded fact that this information was already covered and the SSA is sick to death of getting messed with.

SSA officials said, in keeping with its mission, the agency is primarily focused on ensuring that it does not make benefit payments to deceased Social Security program beneficiaries. As a result, it only verifies death reports received for individuals who are current program beneficiaries, and even then, only for those reports received from sources it considers to be less accurate.[43]

For example, SSA officials consider death reports from states that have a pre-verified decedent's' name and Social Security number to be fairly accurate, so the SSA does not verify that the subjects of these reports are actually deceased. It would, however, verify a report received from a source such as a post office. The SSA verifies no death reports for individuals who are not beneficiaries, regardless of source. Because there are a number of death reports that the SSA does not verify, the agency risks including incorrect death information in the Death Master File, such as including living individuals in the file or not including deceased individuals. Specifically, for death reports that are not verified, the SSA would not know with certainty if the individuals are correctly reported as dead.

With some sympathy for the SSA established, let us examine what is so wrong with their lack of verification. The

SSA does not verify death data of non-beneficiaries. This may have been acceptable when only the SSA was using their data for their own purposes, but that is not the case any more, so the verification standards should be reevaluated. Non-beneficiaries are our first branch of unverified death data. After those data points (referring to the deceased as data points, I know, it's grim), how or who reports the death to the SSA determines whether or not it is verified.

Whether the SSA verifies death reports depends upon (A) whether the decedent is receiving Social Security benefits, and (B) the source of the report. Verification includes confirming the date of death and decedent's Social Security number to ensure that the person identified in the death report is the person who died. Sometimes, a simple error can result in a person who is still alive being listed on this file, meaning they would run into problems whenever they attempted to file taxes or apply for benefits, or even simply apply for a credit card. If someone ran their identity through the Partial Death Master File, they would appear as dead, and look like they were the one committing fraud.

The SSA only verifies death reports for individuals currently receiving Social Security benefits. Even then, the SSA verifies only those reports from sources it considers to be less accurate, such as financial institutions and federal agencies. Funeral directors and family members, on the other hand, are considered accurate sources. Deaths reported from these parties are often not verified, and go straight to the Numident. The following examples of how bad death data can still be included on the Numident come from the GAO's number one best seller, "Preliminary Observations on the Death Master File," my personal favorite. Truly gripping stuff. I am not kidding. Do not read it before bed if you have to wake up early. I rewrote the cases to make them a tad more accessible.

- Steve gets a Social Security benefit check. A post office returns that check to the SSA noting that Steve is now dead. Poor Steve. Since Steve's death is not reported by Steve's family, but the post office, which is a less accurate source, it is turned over to an SSA field office to verify. Once the field office reaches out to Steve's family to confirm his death, he is inscribed in the Numident, and consequently, the Full and Partial Death Master Files.
- Veterans Affairs submits Lisa's death report to the SSA. The SSA determines Lisa is not receiving Social Security benefits. The SSA does not verify the death before recording it in the Numident.
    - Do you see how this is problematic? Lisa could not be receiving Social Security benefits for multiple different reasons. It is not a guarantee that she is dead.

If a death is reported through the Electronic Death Registration System (EDRS), a program developed to transmit death data faster and with more accuracy, or from a funeral director or family member, that death is not actually verified. According to a May 2013 GAO report, "Because there are a number of death reports that the SSA does not verify, the agency risks including incorrect death information in the Death Master File, such as including living individuals in the file or not including deceased individuals."[44] Unverified death data is contributing to why these databases are not reliable.

Additionally, there are problems within the data set based on matching information:

> The SSA also does not record some deaths because incorrect or incomplete information included in death reports generally prevents the SSA from matching decedents to the SSA records. For example, if the SSA is unable to match a death report to data in its records such as name and Social Security Number (SSN), it generally does not follow up to correct the non-match and does not record the death.[45]

There are so many different ways for information to get entered incorrectly. It could be on the front end when the report is made, or by family or a funeral director—getting one of several pieces of information wrong. If I went to order a driver's license under the guise of Heather—and her data was entered incorrectly by a number of different parties, the SSA would not have recorded her death on the Numident. This is one place where it is difficult to forgive the organization's poor record keeping, despite this never being their intention. If a reported record does not match up with the data they have, they will simply ignore it.

It would be like if you ordered a burger at a restaurant and they said, "No sorry, we don't have burgers," and then left you sitting at your table to go watch a movie. Instead of offering you something else, or trying to find something similar to a burger, or even recommending another restaurant, they would just leave their job. Ignoring death data that does not initially match up does not make that problem go away, in fact, it probably points to a bigger problem. A very reassuring footnote does tell us, in not too much detail, that the SSA "will take some internal steps to identify the decedent."[46]

Our beloved GAO reports address that the SSA does not keep track of how many deaths are unverified, but due to the many avenues where errors occur, we believe the numbers to be quite high—higher perhaps than the estimated amount. This is a trend that abounds all around the fraud universe: reported numbers do not reflect the much higher numbers of fraud that are occurring.

The SSA's methods for processing death reports may result in inaccurate, incomplete, or untimely information.[47] Using inaccurate, incomplete, and untimely death data can lead to improper payments. The specific procedures include (A) verifying a limited portion of death reports, (B) not including death reports that do not match with the Numident file, and (C) not performing additional reviews of death reports that occurred years or decades in the past. This last point is especially poignant, because identities do not go bad. If anything, they gain value over time. If the SSA does not retroactively check their old death data, which they know is more likely to be inaccurate due to physical processing, those identities are still fair game.

The GAO analyzed erroneous death data in the Full Death Master File. The conclusion? If the SSA had verified the death reports when it received them, they would not have happened. Here is the hard data. The GAO identified nearly 8,200 deaths that the SSA deleted from its death data between February 2012 and January 2013. Meaning, a death report matched a record in the Numident, the SSA marked it as deceased, then later unmarked it. Apparently, this occurs because the decedent turned out to be alive or was misidentified as another individual, or as a result of plain old-data entry errors.

The GAO drew 46 random cases out of the 8,200 and asked the SSA to figure out why they were randomly deleted. In 28 out of 46 cases, the SSA was able to figure out why they were deleted: 12 were false death (so the person was still alive), and four occurred because identifying information for the wrong person was included. For the remaining 18 of 46 cases, the SSA "was unable to determine from its records the reason for deleting the case from its death data." So for the almost 40 percent of cases where failure occurred, we do not even know why it happened. Nine of the 46 cases involved non-beneficiaries, which are not verified, four of which the SSA was unable to determine the reason for deleting the cases from its death data.

If you read this section carefully, you will notice that these numbers do not add up. Twelve of the 28 cases in which the cause of data deletion was known simply are not mentioned—and then we have to parse which instances (two of the 13 out of 46 reported by families they do not know about, and four out of the nine of 46 who were non-beneficiaries they do not know about it). I made charts, and frankly, it still does not make much sense. Some of these numbers are just left out.

According to the GAO, "most of these errors would not have occurred if the SSA had verified the death reports when it received them." You[48] cannot get much more explicit than that. The two overwhelming conclusions I draw from this is that government data needs to be more accessible and well-thought out when it is presented in statistical form, and that the SSA is struggling to get its arms around its own data. By not verifying their data, they are shooting themselves in the foot. Better yet, the SSA keeps track of neither how many death reports it verifies, nor how long these it takes when they do actually verify something.[49]

Moving Forward

The end game of why all this matters—why death data should be accessible to agencies, complete, timely, and verified—is access. Access, or enablement, whatever you want to call it, is what happens when the wrong people have access to government money or documents that do not belong to them. If a DMV cannot accurately verify identities, it has no way to prevent real driver's licenses issued to fraudulent identities. If SNAP benefit offices cannot accurately verify identities, they may pay benefits to people who do not need them or qualify for them. Getting lost in the data is overwhelming, but we cannot afford not to care.

In 2014, improper payments were estimated at $124.7 billion, up about $19 billion from the previous year.[50] These estimates, supplied by the agencies, tend to be conservative. No one wants to be at fault for giving away taxpayer money, or in the GAO's own language, "[we] identified the federal government's inability to determine the full extent to which improper payments occur and reasonably assure that appropriate actions are taken to reduce them as a material weakness in internal control."[51] That is a very nice way of saying that the government cannot find its way out of a paper bag.

Examples of improper payments, due to missing information on death databases are as follows:

- Disaster assistance: Hurricane Sandy disaster assistance sent to 45 identities that appear on the Numident and the Full Death Master File, but not on the Partial Death Master File.
- Farm programs. Despite already telling the USDA that it needed to do a better job of preventing improper payments in June 2013, their process for reviewing outgoing payments was still deemed unacceptable years later.[52]

These programs and others, such as Rural Housing Assistance, have started at least moving in the direction of better reviewing their death data, whether that means checking the Partial Death Master File at all, or clamoring for access to the Full Death Master File. While part of the problem remains that even if they do verify their data against the Partial Death Master File, they are still not receiving the best data, the other part is that they are only beginning to verify data as it comes to them. If someone died in 2015, for example, their death data would most likely be verified. However, if someone died in 2000, or as late as 2012, no one is retroactively verifying those deaths.

Those identities could continue to be fraudulently used to receive benefits, even though the actual recipient is dead. Since no one is verifying older death data, all the identities still in fraudulent use will continue to be exploited.[53] It is problems like this that need to be looked at from a different angle. Sure, verifying from here on out will prevent it from happening in the future, but all the previous identities are not going bad, especially if no one is actually verifying them from the Partial Death Master File.

All of this is terrible, people are losing money, criminals

are descending, Batman is nowhere to be found, etc. etc. So what is being done about this? As of 2013, the SSA had been discussing several initiatives to better their data quality and distribution. Since fraud is an ever-changing game, we think it a good strategy to take a longitudinal view of policy in order to help create better policy in the future.

- Check whether birth dates are before death dates. This seems like a no-brainer. Unfortunately, it was left undecided as to whether or not the SSA would actually correct the dates on incorrect records from before the check was implemented. Kind of defeats the purpose.
- Data matching: in December 2012, the SSA started using other databases such to corroborate death data. If the Numident or other databases showed they were deceased or over 115 years old, they would cancel benefits. They found about 17,000 such cases where they had been sending benefits to the deceased because they simply were not verified.
- Data exchange: in September 2013, the SSA began a data exchange to identify people age 115 and up who were receiving benefits but had not used Medicare recently and had no insurance or nursing home information on file. About 18,600 cases were referred to the SSA.
- The SSA also plans to introduce a computer code that can be used to terminate benefits for 115 and up whose benefits have been continuously suspended for seven or more years, and for whom the SSA does not have a death record.[54]

In addition to giving a longitudinal view and understanding of this data, using the information from 2013 allows us a poignant conclusion. In March 2015, and then later in June 2016, the GAO put out reports that used the lovely and defeating phrase, "As we previously reported," which usually precedes a brief recollection of what the former GAO report said. We find this most often is happening because it is the easiest way to summarize something. This feels basic, but that is because it is. If the GAO, in 2015, has to summarize something from 2013 that was being reported on back in 2001, that tells us that the problem still has not been resolved. It tells us that the GAO has to constantly repeat its recommendations to an exhausting degree, unless they copy and paste one report to the next. In some sections it does appear that they do this, but who can blame them? If nothing has changed, why reinvent the wheel?

From that March 2015 report, the GAO brings up the idea of implementing preventative controls to help avoid government payments going to the wrong people. Preventative controls and front-end measures all translate to essentially the same thing: put a security measure up front so fewer of these fraudulent payments go through in the first place. It is the equivalent of locking your front door versus chasing after someone who just robbed you. If you lock your door, sure, you could still get robbed, but the odds of it just went way down. Whether you can catch up to the thief depends on several variables: are you fast enough to catch them, are they carrying your TV thus slowing them down, are everyone's shoes tied?

That is what we talk about when we say that these kinds of measures also help avoid the "pay and chase" model of recovering money. The "pay and chase" model is wildly inefficient. If someone receives a wrongful payment from the government, odds are low that they are going to call up the SSA or any other benefits agency and demand they take their money back. The GAO's recommendation remains in sharing death data as a way of front-ending this problem, so payments go to the right people in the first place. Cross-referencing data from different sources is also a good call. In many of the previous examples of incorrect death data in the Numident and Death Master Files, these problems were found by different databases containing those individuals' death data, such as Medicare.

Front-end controls are beginning to be implemented, but they still have holes. For example, the Do Not Pay initiative, a data-matching service that allows agencies to check death data before they pay out benefits, except it only allows access to the Partial Death Master File rather than the Full Death Master File. Whether agencies and institutions will have access to the Full Death Master File or not is dependent on legislative change, but where these groups get their Partial Death Master File from is not.

While without movement from the Federal government, nothing can be done about issues of Completeness and Verification within the Partial Death Master File, Timeliness is one factor that can be dealt with. Using commercial data services, "likely costs their agencies less than if they attempted to provide a similarly extensive data service using only agency resources." Using a data service like this could mean that not only is data obtained faster and cheaper, but more reliably as well, and could prevent all the more improper payments for it.[55]

Since budgets are always part of the discussion, the IRS noted that using commercial data service providers allowed their employees to focus on their work rather than dealing with in-person verification. This is another illustrative point. As soon as you outsource identifying your beneficiaries, you invite in a slew of problems of actually identifying them. Identity theft occurs far more easily over the phone. But this is just the point: working with problems in fraud is often a case of between a rock and a hard place.

If you lighten the demands on an agency by outsourcing the identifying components, you are inherently weakening the ties between person and identity, but you are also creating a cheaper, more efficient "solution." As with most government problems, there is not usually a right answer, and someone somewhere in some department, is probably going to be put out because of it. As of 2016, the SSA noted that obtaining death data helps them to prevent around $50 million in improper payments each month.[56] Benefits fraud has a sordid history of being swept under the rug, with representatives claiming to have less than one percent of fraud within their agency, but when $50 million a month, that we know of, is being caught, agencies need to wake up.

4

Where Does the Data Go?

We live in a world of finite resources. Every
time a fraudster is paid there are fewer resources
to go around for legitimate beneficiaries.
— "State Secure Identity Practices
and Policies in 2015"

It is easy to get lost in all-encompassing fraud swamp, but remembering to come back to what we know, and how each piece ties to the big picture, helps keep things organized. This is what we know so far. The SSA is great at inventing things like the Social Security number and the Numident to keep their own agency organized. Other government agencies (and banks, hospitals, schools, and even cable companies) are great at borrowing that information and then complaining that it is not complete or that it does not fit their needs, ignoring the fact that it was never intended to be used in those ways.

We also know that the Numident is the list that the SSA created for themselves to use to keep track of their beneficiaries, and because of Freedom of Information Act requests, they made the Full Death Master File as merely an easier way to access death data. This is what nine special federal benefit agencies get to use to check death data to make sure they are not giving money away to just anybody—especially not just dead anybodies.

The Social Security Act created a rule that both mandated the SSA (the agency) share death data with federal benefit agencies, but also restricts the sharing of state-reported data. This created the Partial Death Master File, which is what every other agency and institution uses to check death data. The Partial Death Master File is problematic for several reasons: it is incomplete, it does not disseminate death data in a timely fashion, and its incomplete and delayed information is often unverified. Even worse, the SSA predicts that the percentage of state-reported deaths will proportionally increase over time. This means that fewer death records will be included on the Partial Death Master File as time goes by, unless there is active legislative change that allows state-reported deaths to be included.[57]

Despite the fact that the Partial Death Master File fails to be a credible source, the GAO has repeatedly and exhaustingly told them to shape up, to which the SSA gently reminds us, this was never their job in the first place.

The reason why this matters ties all the way back to the role of documents in identity fraud: if agencies are not able to check on a reliable set of death data, fraudulent payments will be allowed to get paid out, and fraudulent identities will be allowed to interact with agencies and institutions undisturbed. It means Larry could continue to receive New Jersey State Pension payments in his mother's name. It means I could take Heather's personal information and apply for a driver's license, solidifying my picture with her name on a REAL ID compliant license.

The communication breakdown between federal agencies,

state agencies, and local agencies, like vital records offices, creates opportunity for fraud. If a private citizen can find out about a stranger's death through an online obituary before that state's vital records office, it means someone's entire identity is up for grabs. It essentially allows the deceased's Social Security number to be two steps away from public record. Obtaining their birth certificate allows you to have proof of their (or your) own identity. It is virtually undeniable to an official to have a matching name, Social Security number, birthdate, and backstory, but not necessarily a matching person. Documents are part of the core of identity and they are how we interact with the world—if they cannot be verified, how can we ever be sure of who we are dealing with?

Document fraud, open states, the connection to the SSA's death data—it is all a convoluted mess to get us to where we are. If you have made it this far, you have waded through the bog of uninteresting information to get to the goods. Birth certificates have nothing that should allow them to be used for identifying purposes—no picture ID, no biometrics, nothing—but they are continually used for such. Because of certain states' lack of security when it comes to its citizens' birth and death certificates, these non-identity verifying documents are available to anyone willing to put in the minimal effort.

The lack of good information to double-check identities is minimal and often unavailable in the ways necessary to maintain decent data standards. This is compounded with a lack of communication between states, or boundary blindness, wherein one state has access to necessary information that another state does not have access to. These factors all combine to create an environment for birth and death certificates to become breeder documents, meaning that if you are in possession of one, you are often able to get the other—meaning that if you have a birth certificate, you can get a license or a passport. If you have either of those official identifiers, you can do just about anything.

Washington, being an unabashedly open state, is one of the most forthcoming states concerning information. Just to test the waters, I figured I'd try to order a legitimate birth certificate with all information provided from a known source. My boyfriend was born in Washington state, so with his reluctant permission I entered his full name, date of birth, city of birth, and parents' full names (including his mother's maiden name) into VitalChek, the website where you can order most states' birth certificates online. In the middle of filling out the information for the birth certificate, VitalChek has a disclaimer to let the user know that Washington is an open state and anyone can apply for any birth certificate, as long as they have all of the correct information, or so they say.

Similar to when I applied for my friend's birth certificate, there was some information I was not 100 percent certain about. I had no idea whether his parents' middle names were listed out fully on the birth certificate or not, so I wrote them out. Lo and behold, when the certificate arrived, only the initials were listed, but so was the certificate number, filing number, filing date, and every other thing necessary for a certified, useable birth certificate. I was shocked at how easy it was. However, we wanted to be sure we came at this from all angles in our research. Being able to lean across the table to confirm what city he was born in makes it easier, but identity fraud is not unique from other crimes: a lot of the time, it is perpetrated by someone you know.

Obituaries are one of the most overlooked and underestimated sources of information for identity fraud. All you have to do is look in your local newspaper, or search the terms, 'obituaries' and 'Washington,' as I did. Maybe people don't realize just how much personal information they are revealing when they publish an obituary. Full name (including middle and maiden name), exact date of birth, exact date of death, the city where the individual was born, where they lived, their parents names.

The obituary, by its nature, acts as a tribute to the individual who died by categorically outlining the details of their life. Unfortunately, that detailed tribute provides the exact information someone would need to order a copy of their birth certificate. Rather than use someone's data that I already knew or could obtain easily, I wanted to try to order someone's birth certificate in Washington without knowing a single thing about them, other than what I could find on the internet (and preferably for free).

You have to comb through the obituaries to make sure they have all the information you need. The biggest problem I ran into was figuring out the parents' middle names and the mother's maiden name. So, Washington being an open book, I searched for the parents' marriage record. This is where the ancestry sites came in. These kinds of sites make it really easy to search just about anyone. I signed up for a free trial and instantly had access to all kinds of information. It was, to be honest, overwhelming.

Some of the most depressing things I stumbled upon through this journey with documents were the piss-poor attempts to keep personal information safe. On one document ordering site, it states, "Since your official birth certificate is considered an identity document that is necessary to apply for things like a driver's license or passport, it is important that you keep it out of the hands of others who shouldn't have access to such a critical document." On the same site, they note that it is fairly difficult to obtain someone else's birth certificate. This is untrue. In the same envelope that brought me the birth certificates from Washington, a state as open as it gets, was a letter extolling the virtues of identity freezing services as a means of protecting your identity. Working within the identity sector often feels like trying to fill a bucket of water with a bunch of holes in it: pointless.

Now for the case of Briana's identity, the birth certificate with a giant "DECEASED" typed across the bottom. 20+ years after receiving her death certificate, the vital statistics office finally had time to mark her birth certificate. It is important to note that Briana was born and died in the same state. Of course, when you are looking for an individual, it is a little more complicated to track a death across state lines, but if you are the vital statistics office, it is nearly impossible. The likelihood of a birth certificate being marked as deceased decreases in two aspects: the faster you apply for a birth certificate after learning of a death, the less likely it will be marked, and if the individual was born in died in two separate states, that information will have to go through the Partial Death Master File trickle-down in order to reach the birth state's vital records office.

Heather's Synthetic Identity

My search for an ideal identity began by searching for an obituary, true, but in looking for Heather's parents' maiden names on an ancestry search engine website, I came across something very interesting. The Heather I was familiar with was born in Kentucky in 1989, but right underneath her entry on the ancestry website was another Heather—with the exact same middle and last name, born in the same year, in the same month, five days apart. This Heather was born in Louisiana, and had died in 1999 rather than 2017, and similarly to Briana's entry on the ancestry website, it listed her Social Security number.

So there I was, with two frighteningly similar identities, both deceased, with the exact same name, and one Social Security number. Imagine this as a word problem from elementary school: if you have three different ice cream flavors, and nine flavor toppings, how many combinations of one flavor and two topping combinations can you make?

I have read a lot about synthetic identity fraud at this point, but reading about it and watching the possibilities for it transform in front of you are different. I was looking at an identity theft smorgasbord: how many different ways could I combine these two identities to create the ultimate fraud sundae?

I began by taking the original Heather's identity and combining with the second Heather's Social Security number (at this point in the process, I had not yet obtained the original Heather's death certificate, and thus did not have her Social Security number). This was the first iteration of a synthetic identity—identity fraud that uses a partially made-up identity. Synthetic identity fraud runs the gamut from fully made-up to partially made-up. In this case, all the components were real, and if anyone checked, a name and a Social Security number would match.

After receiving the original Heather's death certificate, I now had two identical names, two unique Social Security numbers, and one useable birth certificate. I did not bother to order the second Heather's, since she died in 1999 and it is likely that her birth certificate would be marked as deceased. But despite not having Heather 2.0's birth certificate, the combinations start to feel dizzying. In this instance, the synthetic identity fraud would be difficult to detect because the two identities look incredibly similar. And knowing what we know about death data and how frequently issues within it are resolved as typos, how likely would it be that a birth date five days off would be overlooked as a mistake?

We cannot go through every iteration of synthetic identity fraud potential, but imagine I created two W-2s—one with Heather 1.0's Social Security number, and another with Heather 2.0's. I have two passable identities, one real, one synthetic. When you start fabricating documents, it is frightening to realize how easy it is to do. Since businesses only verify the amount of money and their EIN on tax returns (and not necessarily the name of the employee), having a not-quite-accurate name would not raise any red flags.

This is where synthetic identity theft blurs the line: all of the documents I created for Heather 1.0 *are* real, they are just edited. I can also use those same documents to verify Heather 2.0. Creating synthetic identities is not nearly the amount of effort people think it is. It is simple, it is elegant, and it is hard to detect. Sussing out the flaws within the identity system is simple. You can track the information, you can test it. And if you are brave enough, you can use it.

At this point, I now have every single document I would need to apply for a Wyoming driver's license, or any driver's license for that matter. Since those are the tenets for applying for a Real ID certified form of identity, you could take those particular documents, apply for a Wyoming driver's license, and then move to another state and surrender your Wyoming

ID for a local version. Heather's death was recent enough that it would not be included on the Partial Death Master File.

This is why documents matter: because they back up an identity. Identities establish your presence, your very existence. If you have a legitimate one, it is as if you have a Get Out of Jail Free card for life. It means if you are a wanted criminal, and you get pulled over for speeding, you are likely to walk away unnoticed. It means you have a shield. It means you can apply for benefits, take out credit cards.

You do not have to just take mine and Larry's word for it. There is a book series called *The Paper Trip.* "The Paper Trip I," the first of the four volume series, was published in 1971 by a disgruntled, anti-government individual as a guide for developing new identities in order to dodge the Vietnam War draft. The book series has grown, focusing on new and emerging techniques to help individuals, for whatever reason, get out of their existing identities and forge new ones.

"The Paper Trip II" focuses on name change. "The Paper Trip III" is all about using birth certificates and Social Security numbers to start life over. The edition that we are focused on, "The Paper Trip IV," explores the complications put in place by the REAL ID Act, and how to circumvent issues with identity through that. There are lots of criminal guidelines out there, but *The Paper Trip* series, while plagued by a conspiracy-theory rhetoric and prone to rants about Big Brother, is very, very thorough.

Seeing all the same research we performed, already published, did make us feel a little bad about ourselves. But putting ego aside, it highlighted one important thing: everyone who is committing document fraud already knows about this.

The only people who do not seem to know are the people trying to stop it.

The REAL ID section of the "Paper Trip IV" is perhaps even more thorough than any government document or website I have seen. This guy did his research. He lists out data requirements, cites exact rulings, discusses the linking of licenses and ID card databases, and highlights each state's position as being for or against the REAL ID Act. As misguided as his notions may be, from one data nerd to another, I have major respect.

The "Ways of Creating New Identity" section of "The Paper Trip IV" lists exactly what we have discussed: using the identities of a young, deceased person, ordering birth certificates, and (something we will get to in our next section) using a business as an identity. The whole second half of the fourth book outlines every single state and their requirements for ordering birth and death certificates. He lists what phone number you need to call, what links you might have to follow for ordering online, and even the exact cost in each state. He talks about the shortfalls of the Social Security number and how there is nothing secure about it.

It is a 217 page guide on exactly how to commit identity fraud. The interesting catch? The author fully disengages from the identity fraud aspect. In an afterword at the end of the book, the author writes:

> It never occurred to us, however, that someone in need of a new, honest start in life would use paper trip methods to pursue a new, but criminal life. The road to perdition was mapped, and the thieves began driving it with abandon. Identity theft became an instant crime of opportunity in the Seventies (and to this day), needing only

> a change-of-address form, easy-to-get credit applications, and a few postage stamps... Identity theft thus became the ugly illegitimate progeny of the paper trip.

The Paper Trip series author even goes so far as to compare a national ID crisis with Nazism, calling it (his caps, not mine) "the End of the Road...". Radicalism aside, he claims no responsibility for handing out the keys to commit identity theft. Whether that is what he intended or not, it is obvious that this information is now there for the taking.

The documents that we put our faith in to safeguard our identities are breeders; they bring us other documentation. Breeder documents, like birth certificates (and more surprisingly, death certificates) lack standardization. The Secure ID Coalition suggests standardizing these breeder documents, and thinking about the mechanisms that are both front-end and back-end. This addresses the question, "what tools are we going to use to stop this?"[58]

The Electronic Death Registration System

The Electronic Death Registration System (EDRS) is exactly what it sounds like. It is an electronic way to register deaths. The EDRS plays a key role in preventing fraud by making the death registration process more efficient, timely, and of higher quality. The EDRS is used in most states, but not all, and has been implemented in different ways and at different times. Although the states differ in their specific processes for reporting deaths in the EDRS, the overall process involves the funeral director entering the decedent's personal information and signing it, the medical certifier entering the death information and signing it, and then the proper registrar reviewing and registering the record. Each EDRS slightly differs in its benefits, but across the board they all feature increased efficiency, timeliness, and quality. States that use an EDRS now have an easier and more reliable way to register deaths, which has huge implications for administration of law, public health, and preventing fraud all over the country.

Combatting boundary blindness, the inability to see and use data across state lines, is a crucial step in the process of minimizing fraud. The less states communicate, the more fraudsters have to gain. This occurs in every arena of fraud, but particularly when it comes to documents and death data. If I take Heather from Kentucky's information, and obtain a Wyoming driver's license with it, it makes it all that much harder for Wyoming to track Heather's original identity. This means if I applied for benefits under Heather's identity in Wyoming with my Wyoming license, it is more likely that I would actually receive those benefits than it is for Wyoming to figure out I was actually dead. EDRS is hoping to minimize the problems created by boundary blindness and to build a national repository for death data. The internet has an easier time seeing across state lines than paper ever did.

Official death records are governed by state and not federal laws, so these records cannot be released publicly by the federal government.[59] According to a report[60] from the GAO, the SSA "receives death reports from multiple sources, but its procedures for collecting, verifying, and maintaining death reports could result in erroneous or untimely death information." Amending the Social Security Act so the SSA can legally share state-reported death data, requiring the SSA to make sure their death information is as complete as possible, delivered quickly, and gets verified, are all good steps.

However, given the fact that the SSA's death data was not created for keeping track of identities, and given the fact that fraudsters are taking advantage of the country's lack of accurate death data, perhaps the best solution is to create a new system specifically designed for monitoring identities and death data. An electronic system would allow data to be accessible to more entities, it would be easier to keep up to date (and therefore more complete), and it would be easier to verify information. Something like this already exists in most states, the aforementioned EDRS, which is run by the National Association for Public Health Statistics and Information Systems (NAPHSIS).

NAPHSIS is the national nonprofit organization representing the state vital records and public health statistics offices in the United States. Formed in 1933, NAPHSIS brings together more than 250 public health professionals from each state, the five territories, New York City, and the District of Columbia. NAPHSIS has several systems they have implemented to help curb disorganization within the vital records called Electronic Verification of Vital Events (EVVE), State and Territorial Exchange of Vital Events (STEVE), and EVVE Fact of Death (EVVE FOD). These systems act as intermediaries between agencies and the vital records offices of most states and jurisdictions, helping to coordinate death data across the nation and prevent fraud.

The History of EDRS

In the early 2000's, the SSA started funding for EDRS. In 2002, a national model was developed, which was funded by both the SSA and by the National Center for Health Statistics (NCHS). It also had participation from multiple vital records jurisdictions: The National Association for Public Health Statistics and Information Systems (NAPHSIS), NCHS, and the SSA. NAPHSIS played an important role in the creation of EDRS; they developed documentation (e.g. standards and guidelines, training materials, etc.), reached out to national death data provider organizations, facilitated communication, provided consulting services for implementing the EDRS, assisted jurisdictions with modifying the EDRS National Model to their needs, and spoke at jurisdiction EDRS working groups.[61]

In 2005, the SSA planned for the EDRS not to be a brand new collection of death information, but rather to be a replacement of "the cumbersome and labor-intensive process under which the SSA currently receives death information." They said they would disclose the EDRS to "other entities when authorized to do so by Federal law."[62]

The GAO noted in their report[63] on improper payments that, "sharing death data can help prevent improper payments to deceased individuals or those who use deceased individuals' identities," however, death data is quite challenging to obtain and to maintain. The old process of death registration data flow was very time-consuming and allowed much room for error. Traditionally, the process was paper-based, which left a large margin for error.

This general data flow of reporting deaths is complex. It

involves many stakeholders and many shared responsibilities. When a death occurs, a funeral home or medical examiner enters the fact of death and demographic information on the death certificate, and the medical certifier enters the cause of death on the certificate.

Next, also at the local level, a funeral home or medical examiner files the death certificate with the local registrar, and the registrar files the death certificate with the state. At the jurisdiction level, the state vital statistics office sends death data to the national level, unless there were issues with the certificate, in which case it goes back to the local registrar. Lastly, at the national level, the cause of death data is reviewed and/or assigned codes for diseases or abnormal findings. If there are issues with the death certificate information, it gets sent back to the state's local vital statistics office, and if there are no issues, finally gets filed.[64]

As of February 2014, "Every state in this country had either a complete Electronic Death Registration System (EDRS) in place right now, is in the process of formulating one, or is contemplating creating one."[65] It is good to know that at the very least they are thinking about it. Some states have their own personalized names for their EDRS. South Carolina's first EDRS was called WebDeath and Louisiana calls theirs the Louisiana Electronic Event Registration System (LEERS). It's hard to know what sounds worse, WebDeath or LEERS, but as bad as the names are, at least they have one.

Connecticut, North Carolina, Rhode Island, and West Virginia have no EDRS. Rhode Island does not even have a plan for one because it is considered a closed state. If you recall, Massachusetts is also a quote-unquote closed state, and a state I have multiple birth certificates from. According to

VitalChek, "Rhode Island death certificates may be ordered online for the person named on the certificate by a member of his/her immediate family," and just like in Massachusetts, this is not set in stone.

They ask the basic questions about the individual: name, place of death, and date of death. I then put in my name and select my relationship to the deceased from a drop-down menu. How could they possibly know I was not someone's daughter or sister or cousin who has a different last name? What if I just lie and change my name and then put my address? It is ludicrous for the security of vital records to hinge on a drop-down menu.

Not every state is required to have an EDRS, and just because one exists does not guarantee a more accurate set of death data. You actually have to use it for it to be useful. According to NAPHSIS, "states that mandate the reporting of deaths through EDRS generally should have a higher percentage of EDRS reporting."[66] It is a strong "should" in that sentence, and a telling one. It's as effective as having a seatbelt and not wearing it, or in this case, having seatbelts and no law requiring anyone to actually use it. The states that don't mandate their EDRS are letting all that good technology go to waste. Agencies find it inconvenient and expensive to train their employees to use a new system, despite the fact that using an EDRS could save unimaginable amounts of money by preventing fraudulent payments.

Implementing and Using EDRS

Implementing a complex system like an EDRS is no easy task; many users are working on the same death record, the system requires extensive edit checks, and high levels of security. It is expensive to develop, it takes time to customize the system to the requirements of each state, and many people in different industries need to be trained to use the system. Systems problems forced Connecticut to end its EDRS rollout in 2012,[67] but they are still in the process of developing a new one.

There is a wide range of dates for when the EDRS was implemented in each state. Minnesota's first EDRS went into effect as early as 1997, while Colorado and Wyoming's EDRS and were not fully implemented until 2015. Nevada did not get up to speed with an EDRS until 2016. These systems must support national standards and guidelines while also accommodating the laws, regulations, and business practices of individual states, yet another reason to delay implementing one.[68]

States have used different processes for implementing their EDRS, such as initial trial runs and multiple phases of implementation. South Carolina's EDRS first deployed in 2016, and currently, a second electronic system for registration of vital events including death is being updated. In Tennessee, all funeral homes will be on the EDRS, but only some physicians are currently trained for it. Allowing the individual responsible for confirming the death to report it will help streamline the process.

California continually tweaks their system as they learn what is helpful or not. They created a training environment for EDRS and allowed physicians to remotely attest medical information by voice-signature or fax-signature. Good connections with hospital medical records staff and risk management staff helped carry the word of the EDRS initially, continuing to get new providers registered with the system. Having a forward-thinking program that is adaptable will certainly get people more excited to work with it than one that refuses to change to different scenarios.

The EDRS process is similar to the traditional paper-based death registration process from when a death occurs to when a death is registered. Typically, the funeral director starts the record and assigns the certifier. The certifier can then go in, list the cause of death and sign the record. Once signed by the funeral director and certifier, the record is filed with the proper registrar. Most states follow this general process.

The funeral directors, generally being the first to initiate the death registration process, collect personal and demographic information obtained from the family about the deceased. If a Social Security number is included, the EDRS can provide an online, real-time verification that the name and number match. You can understand why that would be important.

The Benefits of EDRS

Although the states differ slightly in their EDRS benefits and values, they are still fairly consistent. The main benefits of the EDRS include increased efficiency, timeliness, and quality. These benefits are a huge step toward fraud prevention; specifically, by having an automated means to mark birth records with deceased markers, these systems make it more difficult for fraudsters to steal the identities of deceased individuals.[69]

Similarly, "the rapid matching of birth and death certificates, a feature of the system, will help to guard against identity theft."[70] A study was even done to evaluate New York City's EDRS in conducting mortality surveillance during and after Hurricane Sandy. After analyzing death records from 2010 to

2012 and assessing the system's components, the researchers found that despite widespread disruptions, the EDRS was stable, timely, and successful in providing causes of death.[71]

EDRS can also assist in surveillance, since analysis of death data in real time is now possible with these systems. Electronic reporting eliminates the month-long, and sometimes year-long, gap between when a death occurs and when information about that death is shared with stakeholders.

Since the EDRS serves as the most common method for reporting deaths, it helps with the collection and interpretation of data at the national and international level,[72] meaning that boundary blindness is less of a concern. It also allows immediate access to revise information on the death certificate at any time prior to the local registrar's acceptance of the form, meaning errors can be corrected quickly. The Vermont Health Department has explained that "The EDRS has increased efficiency in the death reporting process by reducing the time it takes to finalize a death record from 38 days to just four days."

Washington State's Department of Health notes that in replacing the paper process previously used for filing death records, the new electronic filing greatly diminishes the time it takes to receive death certificates, and it also improves the quality of the data on causes of death.[73] Similarly, Kentucky's Department for Public Health highlights some of the key features of the EDRS as saving time and effort, improving turnaround time for obtaining certified copies, and allowing error correction and enhanced accuracy.[74] Both Washington and Kentucky are problematic states, seeing as I have birth certificates for a deceased individual from both of them that are not marked as such. As much as I want to trust that these

EDRS programs are greatly reducing fraud, I have very recently obtained certificates that prove otherwise.

Another remarkable feature of the EDRS is its ability to reduce the rejection of death certificates (unlike the SSA, which when faced with an incorrect data entry, will reject it and not record it on the Partial Death Master File rather than resolving the issue). The EDRS "checks automatically for many types of errors and prevents unauthorized access or alterations to the information."[75]

Along the lines of timeliness, the filing of death records electronically must occur within the required 72 hours after a death occurs.[76] Mandating a faster response time for records to be recorded online is a really important step. Overall, improved communication among these entities allows for more accurate and reliable information in the database. The states that don't have an EDRS logically don't reap the same benefits. Thus, they are affected by the paper-based process of registering deaths in ways that are much more time-consuming and vulnerable to errors.

Have EDRS Systems Actually Fixed Anything?

Since Georgia launched their EDRS, they have seen increased demand for their death data. They regularly provide electronic death data to Georgia Department of Public Health programs, other state agencies, and researchers. Additionally, Florida sends death data out at the end of each day, giving their EDRS partners a daily input. They also verify Social Security numbers on a regular basis, allowing the SSA to be notified in real time so they can flag any pertinent data.

This faster death reporting due to the EDRS has implications for public health, for example, during influenza outbreaks. Idaho's EDRS has a notification set up for anytime an influenza death is reported. Since influenza outbreaks are short-lived, relying on paper death certificates means that by the time the scope of an influenza outbreak is known to the State Health Officer, any preventative or educational efforts would be too late because the outbreak would have already peaked. Alaska has also felt the benefits because they are now able to report deaths on a weekly basis, whereas before they reported deaths monthly.

Data from the Office of the Inspector General showed that from 2012 to 2014, the SSA received 41 percent of EDRS reports within five days of death. That is a significant improvement from previous years, when the average was 11 or 13 days. The extra six to eight days may not sound like a lot, until you imagine how much fraud can slip through the cracks. If I were to apply for a fraudulent driver's license, it would only take 40 minutes.

Even though the EDRS had been implemented by many states, those states weren't reporting all of their deaths using the system,[77] meaning the offending states are still not using their better death data to its best advantage. Reporting more (and ideally all) deaths to the SSA would be better use of the EDRS to ensure that the SSA data remains up-to-date, complete, and accurate. After all, the SSA's death data is what informs most of our government agencies.

EVVE and EVVE FOD

EVVE, created in 2002, is a national electronic system operated by NAPHSIS that allows its customers to quickly, reliably, and securely verify and certify birth and death information. It serves as a middleman between its users and the Vital Records Offices around the country. EVVE's customers include both federal agencies and state agencies such as the SSA, the Department of Motor Vehicles, Medicaid offices, the Army National Guard, regional FBI offices, and the Department of Homeland Security.[78] Electronic inquiries from authorized users can be matched against over 250 million birth and death records from state- and jurisdiction-owned vital record databases all over the country.

An electronic response from the state or jurisdiction either verifies or denies a match within seconds. It is the only system that provides access to the most complete set of state and jurisdiction owned vital records.[79] Most states use an EDRS, so when a customer is searching for death data using EVVE, EVVE is checking each system to quickly obtain death information.

With EVVE, a government office anywhere in the country can get immediate confirmation of the legitimacy of an applicant's United States birth certificate. EVVE's authorized customers, as described above, can send an electronic request to any participating vital records jurisdiction, either to verify the contents of a paper birth certificate, or to request an electronic certification. The participating vital records jurisdiction either verifies or denies the match with the death data found within their EDRS.[80]

The EVVE system will also flag responses in which the

person matched is actually deceased, which is a huge step toward preventing fraud.[81] A passport service issuing a passport will run the person's birth certificate through EVVE. They enter the person's information, such as the person's name, date of birth, and other information from his or her birth certificate. They get a match, and on the file it says, "Living Indicator: No." In this instance, the passport service would know they may have a case of fraud on their hands.

It generally takes a while for deceased individuals' birth certificates to get marked as deceased, but a system called State and Territorial Exchange of Vital Events (STEVE) is trying to change that. STEVE acts as a much needed third party to help send death information across state lines, where it can typically get lost. STEVE allows states to quickly and accurately coordinate with one another.

EVVE FOD was created in 2017 and informs both government and private agencies if the people searched in their system are deceased. When using EVVE FOD, the user does not need to know the time of death, date of death, and even if a death has definitively occurred. Authorized EVVE FOD users include insurance companies, financial organizations, federal benefit organizations, and pension groups.[82]

According to NAPHSIS, "ever since the [Full] Death Master File [thus creating the Partial Death Master File] removed state protected records from its database, there has been strong demand for a new system that contains accurate death information. EVVE FOD is the answer."[83] The death databases of vital records offices contain the most accurate death data in the nation because these official certificates are the only full sources of birth and death information: "they are the 'gold standard,' providing the most accurate, reliable, and complete information about death."[84] The only places that create and hold those birth and death certificates are vital records offices, meaning that those are the places we must secure if we want to secure this data. EVVE FOD allows us, as best we can, to do that.

Currently, 41 states (63 percent of the death data in the country) are on EVVE FOD. EVVE FOD works by matching death records from a credentialed user against databases of vital records offices all over the country. The user can send out one record, or millions of records, and will receive matches within seconds, finding out who among the records submitted is deceased, as well as their places of death and dates of death.[85] The user's request gets sent to every state's and jurisdiction's EDRS, making it one big collaborative effort.

The problems at hand are that the death registration process leaves holes open for fraudsters to creep into, and that government and private agencies do not have access to complete and accurate death data. Implementing an EDRS, which most of the states have already done, helps clean up the process of death registration by making it more efficient, timely, and of higher quality, thus helping prevent fraud. At the national level, EVVE and EVVE FOD are providing complete and accurate death data to the agencies that need it. It seems a way to alleviate these problems is for every state and jurisdiction in the country to implement an EDRS, and to actually use it to its fullest advantage.

EDRS is a wonderful front-end solution. If it is implemented and used, a good portion of fraudulent ventures can be put to bed. Representative Devereux from Vermont, in our conversations about H. 111 and their tightening of access to birth and death certificates, said that his state loves their EDRS. While the implementation of an EDRS is certainly a win, and is seeing returns in the way of non-fraudulent enterprises, there is not a lot of room for complacency.

One of the biggest concerns with regard to document fraud is that boundary blindness allows fraudsters to move across state lines with little concern for being caught. This is second only to the threat of fraudulent birth or death certificate requests made quickly after a death occurs. Washington's EDRS was implemented in 2005, but when I was ordering birth certificates in the winter of 2016, I was still able to receive an unmarked birth certificate for a woman who died months previously.

Kentucky talks about how their EDRS allows for a quicker turn-around for documents, yet somehow, when I tried to obtain Heather's death certificate, she said Vital Records would not have access to it for at least three months. At some point there is a disconnect, and discovering the holes yourself is the only way to verify what all the government reports say. EDRS is a great solution, but clearly, there are still problems.

The natural questions that follow this section focus around one idea: how often is this happening? Are people aware of the problems that accompany documents and their availability? A man in Texas escaped from a halfway house in 1996, after serving time for indecency with a child. For two decades, he lived under the identity of a deceased infant, whose identity he found in a cemetery by looking for a birthdate close to his, the same tactic I used by trolling through online obituaries.

It was an easy thing for the man to acquire a copy of the infant's birth certificate, and then use it to apply for a Social Security number in the infant's name. He lived under that guise for twenty years, in Texas, Mississippi, Tennessee, and Pennsylvania. This fits well with our research. Due to the lack of communication and data sharing between states, it makes sense that this man moved around, since it makes it all the more difficult for his old identity to be known.

His identity began to fall apart not due to a thorough investigation, but because of the deceased infant's aunt, who started working on a family lineage project. Public records had started compiling under her dead nephew's identity, which showed up on an ancestry website. Her perturbation makes sense; you would not expect to see your dead infant nephew linked to marriage records and different residences.

According to court documents, the man's "successful acquisition of [the infant's Social Security number] has allowed him to assume [his] identity, secure employment, obtain credit, open at least one bank account, obtain housing, obtain a professional license, and get married." It[86] is exactly this, the enablement, the question posed by my friend: *what are you going to do with it?* The laws reflect the problems. Possessing sensitive documents is not the problem, it is what you can do with them.

The use of genuine documents to forge a new identity of a different person, or a synthetic identity, creates the kind of enablement for free-range fraud. It can reach into every arena: health care, taxes, benefits, insurance. Every institution and agency has a stake in identity fraud, and a responsibility to not only try to prevent it, but help clean up the damage after it happens. Whether it means taking a hard look at your state's legislation or tightening inter-office practices about safeguarding information, everyone plays a part.[87]

Section II

Business as Usual

5

Business Identity Fraud, or, Identity Fraud on Steroids Wearing a Tie

> Dwight Schrute: You know what? Imitation is the most sincere form of flattery, so I thank you.
> [Jim places a bobblehead doll on his desk.]
> Dwight Schrute: Identity theft is not a joke, Jim. It affects millions of families every year.
> — "Product Recall," The Office, 2007.

Business identity fraud. There is something about the word *business* that just makes you sit up straighter. Meaning no offense, the word itself is a little stodgy, like a starched white collar. Businesses are literally "a company or group of people authorized to act as a single entity (legally a person) and recognized as such in law."[88] A business can mean a giant, glass building on Wall Street with fancy bathrooms, or it can mean a taco truck. Either way, businesses allow a group to come together under a single identity. Maybe two people co-own a taco truck. After becoming a business, they are legally recognized as one taco-loving entity.

Business identity fraud gives an air that we are dealing with an entirely new animal, but this is not the case. Fraudsters use similar, if not the same tactics on businesses that they have always used on individual identities. Everything we have discussed about the sensitivity of documents and the value of personal information applies to businesses as well. The major difference between individual and business identity fraud is that misused business identities typically offer higher financial returns.

Many of the examples and tactics we illustrate here will feel familiar, and that is because they are. You should be able to draw parallels between the former section and this one, so go ahead and give yourself a pat on the back when you do. Just like individual identity fraud, there is a massive range of business identity fraud types. So many different tactics fall under this umbrella, but basically it is the use of a business's identity for fraudulent means.

Say Rachel owns a taco truck. Margaret works part-time for Rachel. Margaret pretends to be Rachel and orders tons of chips and salsa in the name of the taco truck, but has the order shipped to her own personal address, so she alone will get the goods. This is a prime example of business identity fraud.

Maybe Margaret creates a separate bank account for the business, using her own name, into which she can discreetly funnel money in and out. Margaret takes out credit cards in the name of the taco truck, sticking Rachel with mysterious bills. That would also be considered business identity fraud.

The combinations are endless. What if Rachel, the taco truck owner, is the fraudster? Maybe she makes up a bunch of employees to start filing fraudulent tax returns, so not only is it tax refund fraud, but regular old identity fraud, as well.

If you can take out a credit card in Rachel's name, why couldn't you take out a credit card in the taco truck's name? Maybe Rachel steals Margaret's identity to manipulate the taco truck business. Maybe the taco truck is not even a real truck, but merely a shell corporation funneling fake taco money to the Cayman Islands.

Maybe Margaret is just a synthetic identity Rachel made up as part of her army of fraudulent employees. At the risk of sounding too meta, I should point out that these are all made-up examples based on real crimes that have happened, but you can see how easy it is to spiral once you get going.

There are thousands of iterations of business identity fraud, just as there are for individual identities. BusinessIDTheft.org calls business identity theft "consumer identity theft's bigger, meaner, and more evil twin." Compounding the two types of identity fraud makes it exponentially more complicated to unravel. Fraud is infinitely creative, and ever-adapting. It is as if a fraudster asked, "If an identity fraud works, why couldn't it work on a bigger scale?"

The Five W's and an H

Anyone who has sat through a high school English class has probably heard of the phrase: *who, what, where, when, why,* and *how*? These basic investigative questions can help us get a handle on what feels like a behemoth of a problem. We have already identified the 'what' section, but the 'who' is a two part question: who is perpetrating fraud, and who is being affected by it? Fraudsters can be anybody. Identity thieves can be people you know, like family members or friends. They can be a random person who goes through your mail looking for a way of accessing credit cards or benefits. They could be a brilliant hacker (less likely, hackers typically just expose information and sell it) or a not-so-brilliant street criminal-turned white collar aficionado. Identity thieves, regardless of whether they are stealing an individual's identity or a business's, can be anybody.

Individual identity fraud has its particular victims: the young, the elderly, and the deceased. What do infants and college kids have in common? Aside from an inability to do their own laundry, odds are they aren't checking their credit scores regularly. The young have clean records and may not check on their credit for a long enough time that it could provide years of unchecked potential for fraudsters. Meanwhile, the elderly have long-established credit and may be less savvy to technological ways of monitoring their scores regularly. While younger victims offer a bigger time frame to steal an identity and not get caught, the elderly have something worth stealing.

Older credit files are not questioned as frequently, and this is another reason the elderly often miss any kind of trouble in their account. That is why Florida, (which had the highest per-person identity theft complaint rate in the nation in 2014 and has remained high on that list ever since) has such significant problems with identity fraud—it is due largely in part to their significant elderly population.[89] And the deceased? They are not checking much of anything, let alone making phone calls to credit bureaus.

If you are stealing an individual's identity, these are your primary targets. But what characteristics do you look for in a business whose identity you want to steal? The most popular victims of business identity theft are small to mid-size businesses. That is not to say that large businesses are exempt, but it occurs less often and is handled in different ways.

Making comparisons with individual identity theft and business identity theft is an interesting task. If you are stealing an infant's identity, they will not have a pre-existing credit file. You will have to create one from scratch and build credit the way you normally would. This may mean taking out a retail credit card with a $50 limit and working up to a larger credit limit. For some fraudsters, this is ideal. It gives them a lot of wiggle room to create a new, healthier identity. Either that, or some quick cash. If your objective is to steal a larger amount of money, but just as quickly, the small to mid-sized business is the equivalent of a college student. According to a Business Week study done in 2007, the typical business credit fraud was 10 times larger than consumer credit fraud.[90]

Smaller businesses can range in credit history, but for a lot of them, constantly monitoring their own credit is not high on their priority list. Small business owners have a lot to handle, and often, monitoring their credit falls by the wayside. Credit levels for small to mid-size businesses can vary, but they are almost always higher than the credit level of an individual. A small to mid-sized business allows fraudsters to get a decent amount of money or product quickly.

Larger businesses, while sharing some similarities, differ in how they operate within the fraudulent realm. Think of the bigger businesses you have heard about in the news. Their names usually share scary headlines with phrases like 'data breach,' 'cyber security,' and most likely either 'Russia' or 'China.' Although it is difficult to separate data breaches and hacking from identity fraud, they are not the same thing. Admittedly, there can be overlap in committing these two crimes. For instance, identity fraud might be used in order to perpetrate a data breach, to help keep the hacker's identity a secret. Data breaches expose the information that can later be used to perpetuate individual or business identity fraud.

For these larger businesses, consumer trust is so important to the success of the business that any identity issues are covered up, especially if there is the possibility of a data breach. In this vein, many larger businesses look at fraud as part of the cost of doing business. This is a terrible mindset, especially since while data breaches and identity fraud are not the same thing, one can lead to the other.

Often business identity fraud is referred to as 'corporate' identity fraud. The difference is slight, but important. Corporations are not the only victims; LLCs, nonprofits, school districts, and any other group that operates as a business, can fall victim to this. Each business has sensitive information that could turnover some form of profit to a potential identity thief.

These pieces of sensitive information vary slightly, but the most important piece is the EIN (Employer Identification Number). The EIN is the Social Security number of the business world. The EIN is how the IRS keeps track of the taxes businesses owe, just like Social Security numbers do for individuals. Having a matching EIN and business information, such as a name and address, is the business version of Personal Identifying Information (PII) that is central to committing individual identity fraud.

EINs are treated much the same way as Social Security numbers, as in, they should be kept private but often aren't. Just like our stack of birth certificates, we have a similar number of EINs we have accumulated through our research. One particular accountant I spoke with echoed the sentiments of other industry experts when he said that business identity fraud would be difficult to commit because I would not be able to find a business's EIN. Also like a Social Security number, they are readily available. If you open your eyes and look around, sensitive information is everywhere.

I once worked as a pottery teacher in a small town art center. One Saturday a month, I would work at the front desk. To get this job, I was "interviewed" (real loose interpretation of interview, meaning I was asked a handful of questions about my experience and nothing else). In the upstairs office, which did not even have a door on it, was an unlocked filing cabinet that contained every single W-2 filled out by every employee the organization ever hired. Social Security numbers, names, and birthdates aside (hundreds of thousands of dollars in potential fraudulent revenue), the business's EIN was on full display.

A lot of this information is either available for free or can be legally purchased. In most states, businesses are required to post their initial and annual filing documents on their Secretary of State's website—documents which contain their business's identifiers. A sales tax number, business license number, all kinds of things that start coming together to amass an identity. This is inherently similar to individual identity theft. The accruing of documents or personally sensitive information in order to bolster an identity. Same tactic, different target.

When business identity fraud happens, the 'what' of the equation will also look similar to that of individual identity fraud. Credit cards can be taken out in that business's name, goods and services can be ordered, and new accounts can be opened. Also like individual identity fraud, businesses and people can hide behind these fraudulent identities. It is less in face than in name, but the idea is the same. Hiding behind a legitimate looking identity, business or individual, gives you access to just about anything.

As I mentioned, business identity fraud is just as convoluted as individual identity fraud. The ways of doing it are innumerable. Tax refund fraud on the individual spectrum is well-known, but it is the same rule of thumb: take what you can do with an individual's identity, and multiply it by a business's assets. You can commit tax refund fraud with a business just as you would with an individual's tax information—though instead of using a Social Security number, you will need the business's EIN.

In 2012, an Atlanta seafood restaurant chain had their identity hijacked. A fraudster obtained the business's EIN and created more than 100 fake W-2 forms, reporting over $4 million in non-existent salaries. The business owner was then slapped with $800,000 in unpaid payroll taxes.[91] Many cases just like this go unreported and losses are swallowed. In the same year, an accountant involved in the case said that such a scam is "very uncommon, but it does happen." Only after the numbers were counted would analysts discover that 2012 was the year with the highest financial fraud losses to date, topping out at $21.5 billion.[92] The trend of stealing business identities happened so quickly that no one realized what was going on for some time before officials took notice. For a national chain, business identity fraud can do some damage. For a small business? It can mean the end, especially if the business does not have the resources to fight fraudulent claims.

Since EINs are readily accessible, especially for publicly held companies that list all their filing documents online, they are like the petulant younger sibling of the Social Security number. Take all of the defects of the Social Security number (no actual security, no picture ID, no functioning databases to keep track of it) and then make it worse. That is the EIN. It took me four minutes to find Whole Foods' EIN online. I'll give you a hint, it starts with a seven.[93]

In 2013, the Treasury Inspector General for Tax Administration (TIGTA) estimated that "the IRS could issue almost $2.3 billion in potentially fraudulent tax refunds based on stolen or falsely obtained EINs each year," and predicted a loss of $11.4 billion over the next five years. The IRS can pick out fake EINs, so if you input a random string of numbers you won't get very far. Having a legitimate EIN makes filing fraudulent tax returns much more likely to pay the refund.

The reason the IRS has trouble matching income and tax returns is because of a time delay. The IRS does not start matching W-2 data until after January 31st deadline. BusinessIDtheft.com says criminals are aware of the delay and act accordingly:

> criminals commonly file their fraudulent returns as early as possible, with the expectation that it could be quite some time before the IRS actually has an opportunity to compare the reported wage and income information to the actual wage and income information reported by the employer. By that time, the fraudulent individual tax return has already been processed, and the fraudulent refund has already been received.[94]

Hypothetically, I could take Heather's information, use a legitimate EIN, like the one from Whole Foods, submit a realistic wage amount, and receive a refund. If I keep the return request low, not only is it likely to go through, but it is likely I would not be caught. The effort it would take law enforcement to track me down, level charges, and prosecute would cost more than I would steal. The cost analysis is simple, and fully in favor of the fraudsters.

'Where' these types of fraud occur is just as complicated. There are examples of fraudsters setting up real-life shell companies, where the physical address is an empty office with a single person answering the phone, to give the appearance of a real business. The 'where' can be in a skyscraper, in a rented suite next to an established business, with packages and credit cards for that business delivered just next door. This type of fraud is called 'address mirroring.'[95]

Most delivery technologies acknowledge an exact match for an address as long as the first line (street name and number) and last line (zip code) match, without necessarily checking the middle bit including a suite, room, or apartment number. Business identity fraud can happen anywhere. It can be online, with no real-life footprint besides an IP address for a computer, through the mail, over the phone, or through the Secretary of State's website. The important thing to remember, always, is that identity fraud, business or otherwise, is only limited by the imagination.

There are lots of reasons why businesses are being targeted either in addition to or over individuals. The main reason is the larger amounts of money up for grabs when a business is involved. Businesses routinely maintain larger bank account balances than consumers, and they also routinely make larger payments. This means the amount of money a fraudster can steal is higher, in general, than with an individual using the same tactics. Generally, businesses deal in larger amounts of money and more goods in every arena than an individual would.

Meaning, a business ordering 100 computers is more believable than an individual ordering 100 computers. If I fraudulently steal an identity, business or individual, those funds can come to be at my disposal, so why wouldn't I aim for the highest amount I can get?

Since businesses usually make bigger and more frequent purchases than individuals, they typically have higher credit limits. Additionally, their credit reports are more readily available because they are intended to show how trustworthy the business is, in the hopes that other businesses will want to work with them and customers will want to buy from them. Many businesses have one central account that additional credit lines can be opened under, meaning that an additional approval process is unnecessary. In other words, credit lines can be opened more quickly and with fewer security checks.

Small and mid-size businesses are more at risk because of that lack of security. While larger businesses often have bigger systems in place to help detect and stop fraudulent transactions, small and mid-size businesses may not have the technology in place to protect themselves.

Just like individual identity fraud, exactly who you target and how is dependent on your end goals: what do you want to get out of the business you are defrauding? Are you looking to funnel money? Make quick cash? Acquire hard goods? Whatever scheme you choose depends on what you want to get out of it.

Business identity fraud capers are just as elaborate and variable as individual identity theft schemes. In one example from 2016, a man was accused of impersonating a McDonald's corporate compliance officer. The man would wear a McDonald's dress shirt and carry some identity-proving documents. He would enter restaurants, get access to secure areas, and steal documents he would later use to get access to hotel rooms, rental cars, and other services—all under the company name.[96]

Sure, this gentleman adapted his own individual identity to take advantage of a business, but what this example highlights so well is that just like individual identity fraud, businesses are at risk when faced with some kind of authority figure. The 16-year old behind the counter is probably not going to question the middle aged man who walks in with confidence, donning a McDonald's corporate shirt and claiming to need access to the back room. A healthy level of skepticism is good in all aspects of life. If you own a business, making sure your employees can think critically about potential fraud is better than the alternative.

This example also tells us that from the business end, more security measures need to be in place. If the accused was able to get access to company benefits such as hotel rooms and rental cars, something along the chain of command did not check out. This individual was able to manipulate the company by using the company's own expectations against them. By matching the description for the kind of person they would expect to call and use those benefits, he was able to slip through the cracks and take advantage.

This is low-level business fraud. He did not steal the business's identity, he pretended to be an actual employee. Let's move up a rung, on the far-reaching, inter-disciplinary ladder of business identity fraud. A man in Iowa was charged with setting up fake businesses, and then using names taken from students on a cultural exchange program to create fake employees. Those "employees" then obtained over $355,000 in unemployment benefits, since the fake business fired the fake employees. According to court documents,[97] the fraudster was later observed on bank video cameras withdrawing cash with the debit cards on which the unemployment benefits were paid.

In this instance, the creation of fake businesses provided a platform to take advantage of government benefits. Business identity theft demands some creativity. Often, these scams are not cut and dry, not clearly one type of fraud or another. This example shows individual identity theft (the students on the exchange program), synthetic business fraud (the creation of fake businesses interacting in real time with agencies and other institutions), and government benefits fraud. Fraud does not just add up; it multiplies exponentially.

Bust-Out Frauds

Difficult as it is to categorize fraud, there is one business identity fraud method that stands out. The typical 'bust-out' fraud uses the pre-established reputation of a business to obtain credit cards, goods and services, and even loans. After establishing some credit through its purchases, the fraudster will often max out their credit limits, 'bust out,' and disappear. This action strands the existing business with debt and angry creditors, much like individual identity fraud. The person or business whose identity was stolen is strapped to their real identity, and the debts accrued by the fraudster, because the fraudster has already evaporated.

Bust-out frauds can also utilize address mirroring by creating a location for their fraudulent operation close to a business they are attempting to defraud, or, by falsifying business registration records online. This allows for the more realistic appearance of a legitimate business that already exists.

Bust-out frauds are already quite lucrative. As early as 2004, estimates for U.S. card issuers were placed at $1.5 billion every year.[98] The credit bureau Experian compiled a list of characteristics to look for in a bust-out fraud:

- An account that is 90 percent or more delinquent
- A balance is close to or over the credit limit (70 percent or higher)
- Checks have bounced
- The account holder cannot be contacted
- This pattern is repeated across accounts at the same or multiple institutions[99]

Bust-outs essentially max out value and disappear. By utilizing the existence of an established business, fraudsters can take advantage of that business's reputation. Bust-out frauds can happen independently or within organized crime circles. This becomes especially dangerous with organized crime. When operating together, bust-out fraudsters can net huge profits. By disguising themselves as trustworthy customers and taking the time to gain trust, either from banks or other businesses, fraudsters look just like everyone else.

One of these organized schemes netted $200 million in credit card fraud. A group of 18 people, ranging in age from 31 to 74, used false identities to bolster credit profiles, and to utilize this undeserved credit to max out cards. This elaborate scheme was operating in eight different countries. They had over 1,800 mailing addresses, and involved the creation of 80 fake companies. One member managed to 'trade' credit histories, adding one credit card's history to a different account.[100]

These crimes are astounding because they allow fraudsters to steal so much, all from behind a computer screen. You don't have to hold up a jewelry store or rob a bank. You simply move numbers around on a screen, and suddenly you can buy a new car. One of the ways to detect these kinds of criminal groups is through targeting their social ties. Some companies are working to create systems that allow businesses and agencies to see how one potential fraudster could lead to the next.

Data Traps

You go to the doctor's office. While begrudging our healthcare system and deductibles, you fill out the forms and peruse old magazines. You briefly wonder why the couch in the waiting room feels moist. You think maybe you do have a latex allergy and you just don't know it yet. You probably aren't thinking about identity security.

Your data, whether it is stored in a filing cabinet or cloud storage, is everywhere. It's at your doctors' offices, schools, and any job you have ever applied for. One of the biggest problems identity security faces could be easily avoided: businesses and agencies collect lots of data, and a good chunk of it is totally unnecessary. When you go to the doctor's office, they do not need your Social Security number. Despite this, most basic forms will ask for it and most patients will give it. That might be a small part of why healthcare has the highest breach incident rate of any industry.[101]

Your health insurance, rather than your individual doctor's office, does need access to your Social Security number. There is no reason why a doctor would need that data of yours, but often, the secretary's online system has a little red asterisk next to the 'Social Security number' box, and if you just told her what it was, she would not have to call Carol over to override the system, and now Carol is involved and asking, "If you could just please cooperate," and promising that, "Ma'am this is a very secure facility, and yes, your concerns about identity theft are very valid but we assure you that our office is perfectly safe, now if you could just give us your Social Sec—ma'am, please do not flip the magazine holder over I just spent half an hour reorganizing it—ma'am, I am afraid I am going to have to ask you to leave."

Take this form for example. I held onto it when I applied for a job at a middle school.

EMPLOYEE INFORMATION

POSITION____________________

LOCATION____________________

ETHNICITY____________________

SOCIAL SECURITY NUMBER____ ____ ____

GENDER: FEMALE____ MALE____

FULL NAME____________________

PREFERRED FIRST NAME IF DIFFERENT FROM FULL NAME____________

MAILING ADDRESS____________________

____

EMAIL ADDRESS____________________

MOBILE PHONE NUMBER (________) ____________

HOME PHONE NUMBER (________) ____________

Yet another miscellaneous form requesting a Social Security number and other personal information, with no real need for it and likely no way to secure it.

I already wrote my Social Security number on the W-2 form for that job, which is the only form you actually need to put it on. Why did I need to write it on a miscellaneous form concerned with my ethnicity? From small to large businesses, schools, and everything in between, the practice of collecting extraneous data is widespread, as is the practice of failing to secure that data.

It is a logical conclusion: the more data you have, the harder it is to secure. Both digital data and hardcopy data, such as hospital forms and W-2s, are at risk. Sean McCleskey is the Director of Organizational Education and Measurement for the Center for Identity at the University of Texas at Austin. Before that, Mr. McCleskey served with the United States Secret Service for 17 years. He recognizes the problems within identity are vast, but as he says, "a lot of the problems come down to too much unnecessary data." That, and the fact that rather than seem unreasonable, people will hand over their information to the Carols of the world willy-nilly.

Mr. McCleskey claims that often, employees and upper-level management are unaware of the information they have. They are unaware of what data is more valuable than other data. How can you protect what you don't know you have, and don't realize is valuable? Employees often cannot answer questions as to why they need particular, sensitive data. According to Mr. McCleskey, "They usually say they have sufficient security without really having a robust program and know the protocol when asked questions about taking, maintaining, and securing data."

Mr. McCleskey told us that "businesses are ripe for the picking." He told us about asking to see a hotel's data security. An employee led him to a back room, and while telling him, "It's all here," looked at the floor where boxes containing customer profiles and other forms should have been, but weren't. Someone had actually come into the business and stolen their physical documents.

"It's lax how they collect their data," Mr. McCleskey said. The best, and simplest practice for secure data collection is this: if you don't need it, don't take it. That way, you have less to protect.

Mr. McCleskey's thoughts mirror our own. His ideas sound an awful lot like a front-end approach: a way of limiting what comes into your hands so it is easier to take care of. This kind of practice is a super easy solution. If doctors' offices do not ask for your Social Security number, no one can steal Social Security numbers from doctors' offices. It is simple, cost-effective, and would really show Carol and her cronies what for.

In most cases, the practice of taking in unnecessary information seems to be grandfathered in with a "this is how it has always been done" approach. Social Security numbers did not originally hold the weight of identity they do now, since so many other agencies have jumped on the bandwagon of using the Social Security number as an identifying factor. The Carols who ask for your Social Security number at the front desk have no idea why their business needs it, only that the form asks for it. By continuing outdated practices, especially those that in the information age have turned harmful, a business is missing the all too important point that data matters.

Dun & Bradstreet and Business Credit

EINs are what the government uses to keep track of a business's taxes, which is why it is likened to a business's Social Security number. A Social Security number, because of its dastardly misuse by other agencies, has multiple functions. An individual's Social Security number keeps track of their taxes and their credit, which makes credit-worthiness a shoddy identifier by proxy.

The Social Security number wears many different hats, which is where the EIN differs. The EIN only keeps track of a business's taxes and can be used as an identifier, but it does not perform any of the other functions that a Social Security number does.

A Dun & Bradstreet number, also called a D-U-N-S number, is run through the company Dun & Bradstreet. It's acquired through an application process and not all businesses have one automatically. It is also not required like an EIN. Essentially, a D-U-N-S number is the non-required credit tracker for a business. To make things even more confusing, EINs and D-U-N-S numbers both contain nine-digits, as does a Social Security number.

While a business is not required to have a D-U-N-S number, they are definitely helpful. If you keep track of your personal credit, you should also keep track of your business's credit. Just like an individual's credit, having a record of your business's credit history can make your business more trustworthy to potential customers. "Why should my grocery chain trust that your taco truck will make good on its payments for crispy taco shells and unidentifiable meat?" "Because we have a D-U-N-S number and can prove that our credit is fantastic, that's why!"[102]

D-U-N-S numbers, while adding another level of security to a business, are by no means a silver bullet in the heart of business fraud. Dun & Bradstreet notes that "more companies are creating shell corporations to disguise business activities and are deliberately misrepresenting their business data to Dun & Bradstreet and government agencies."

This works because Dun & Bradstreet relies on self-reported

data. It's like a resume: if you hustled coffees and licked envelopes, you might glorify your position as 'Executive Assistant.' We all want to make ourselves look better, and so do businesses. If someone is going to check your business's credit information, and you are in charge of reporting it, it's crazy to think that any of that data would be reliable.

A version of this scheme happened in 2013 in Colorado, a leading state for business identity fraud. One man, we'll call him Ben, submitted false documents to the Colorado Secretary of State's website and managed to acquire 12 different businesses. The businesses he acquired had been delinquent, so all he had to do was submit unauthorized statements curing the delinquency and pay a fee. After that, he changed the business address and the registered agent information.

In my own research (and depending on the state), filing the paperwork to reinstate an inactive business or cure a delinquency does not take all that long. The forms are short, and forging a signature is not difficult. The Secretary of State's Office will not follow up with the original owner of the business to be sure it is a legitimate enterprise.

Ben from Colorado also submitted false information to Dun & Bradstreet to increase the creditworthiness of the businesses. After the businesses were established, he sold them. The documents he used to acquire the businesses were fraudulent, meaning he probably forged the signatures of the previous owners. Documents that you can view online, if you want to be extra sneaky, seeing as most businesses have initial filing and annual filing documents as PDFs that are available to the public. These documents include the signatures of the previous owners, so you can even practice manipulating the original signature if you are so inclined.

There are two instances of a problematic lack of verification at play here. The first being that the Secretary of State's website did not verify Ben's documents. A front-end control that could be put into place would require contact from the previous owner giving consent. Of course, a government agency is not inclined to put an expensive verification system in place, but there are certainly ways this could be implemented without it costing a mint. Plus, it is a lot less expensive than dealing with the clean up of a fraud investigation.

The second instance is the fabrication of creditworthiness to Dun & Bradstreet. Dun & Bradstreet operates off self-reported data, meaning that the D-U-N-S number given to each business and the value it holds has a direct relationship to whatever data that business gives them. Self-reported data is inherently problematic. It is like telling a middle schooler they can have ice cream if they got an A on their science project and then asking them what grade they received. If you don't check the grade yourself, you are leaving yourself vulnerable to giving out undeserved ice cream.

Selling the businesses off exponentially increases the returns Ben made; reinstating or reviving a business on the Secretary of State's website generally costs $100 or under, depending on the state. Overall, this scheme was pretty solid. It was uncovered by Colorado's state agencies working together. According to the Chief Deputy Attorney General as published in *The Denver Post*, "While many think of identity theft as a consumer issue, this case exemplifies that businesses are not immune to this crime and need to do their part to stay safe."[103]

An EIN and D-U-N-S number are both important parts to creating a solid identity for a business, though having both does not make a business immune to fraud. According to one Dun

& Bradstreet representative, the compromising or mimicking of an entity's structure is where the fraud happens. Until now, 75 percent of the businesses that cited issues with business identity fraud had 20 or fewer employees, larger businesses are starting to see problems as well. Even colleges and universities are having their identities compromised.

Large business have a lot of numbers running through them, and if their numbers are on par with other orders, often, businesses will not verify who they are dealing with. Businesses seldom report, track, or trend identity thefts. While individuals are quick to report it, businesses often don't like to share external bad news for concern it will hurt their brand, indicate lack of controls, or think that it is not significant enough to report externally. Business to business fraud gets written off as just another cost of doing business, but there are lots real of victims. First, the business doesn't get paid, there is a loss to government in terms of taxes, and costs are passed on to customers who pay higher costs.

One of the bests ways to stay unnoticed is to learn to stay below the radar of often automated credit approval levels. It's difficult to get law enforcement involved when it's relatively small money. It's not as heavily investigated, and you're less likely to be caught and prosecuted. In the end it will be much more lucrative than individual identity theft.

These 'credit levels' are an important aspect to all this. Eric Tamke, the Vice President of Product Deployment & Management and Chief Compliance Officer for Pocket Teller financial services, has worked with Visa Inc., and the United States Department of Defense with secure wireless technologies. Mr. Tamke advised us that if you have a corporate or government credit card in your name, the credit bureaus will not pull a report on your name, but instead for that of the company.

He told us for these kinds of frauds to work, you have to be believable. Ordering things from providers the company has already worked with, and already has an account with is critical. "The idea is to keep it low," says Mr. Tamke. "Submit false payments that can fall in line with other, logical payments that that business would make."

So many of these frauds rely on the same principles that allow document fraud to occur. A D-U-N-S number is a useful identifier, but the problem is all in the self-reporting. I can report a business of a non-existent taco truck, and receive a D-U-N-S number. This number will be used later, by banks or credit card companies, as a way of verifying the existence and non-fraudulent nature of my company.

Shells, Shelves, and Secretaries of State

Synthetic identity fraud is the creation of a new, fraudulent identity from either totally made up information or from pre-existing parts of a different identity. If I used Heather's name with a different Social Security number, that's synthetic. Synthetic identity fraud extends to businesses as well. There are so many different ways to create new identities, business and individual, it is impossible to tabulate every type of fraud. When it comes specifically to synthetic business identity fraud, the most notable, and easily classified, components are when it comes to shell and shelf companies.

Shell and shelf companies are essentially the same thing, but shelf companies are a little older. Both shell and shelf companies are the hollowed-out husk of a company, created for a variety of different reasons, some being perfectly legal. Certain United States tax laws are surprisingly lax, making it legal for shell companies to exist—especially when they are not actually located within our borders. Creating offshore tax havens can be legal, but when shell companies (which conduct no business of their own and contain no real assets) start funneling money to either avoid paying taxes or to create the appearance of a legitimate business for other fraudulent purposes, it starts getting a little dicier.

Shelf companies, named so because they have sat on the shelf longer, are like fine wines and cheeses. They get better with age. Just like with individual identities, business identities gain value over time. Shelf companies are often started by a larger company, and sold off at a later date. Their value accrues over time. If you are an entrepreneur and starting your own company, banks will be more hesitant to give you a loan, but if you purchase a shelf company, giving the appearance of more longevity with a longer credit history, banks will be more forgiving and willing to lend larger amounts.

It is a way of bypassing the checks involved with starting a company by giving the appearance of financial stability. This is not technically illegal, but it has been argued that it is not necessarily moral, and it can be marked on your credit report if it is discovered that your actual, operating business was not the one accruing credit. The credit bureaus can essentially 're-age' your business if they find out you were up to something funny.

There are even businesses for selling businesses. Everyone knows about the Cayman Islands when it comes to questionable business ventures, but back in 2011, Cheyenne, Wyoming had a booming shelf company industry as well. At one address in downtown Cheyenne, over 2,000 companies were registered.

This address, which was filled with mailboxes, a single copy machine, and an employee who answered the phone and took in mail, was the headquarters for Wyoming Corporate Services, which sold shelf companies in addition to other start-up services. While Wyoming Corporate Services and other incorporating groups are not acting against the law, these kinds of businesses enable fraudulent activity. Both state and federal tax authorities have filed liens against companies registered at the address looking to collect more than $300,000 in unpaid taxes.

Among the people who are connected to the single address in Cheyenne are a credit card processor who worked on telemarketing scams, and a former Ukrainian Prime Minister who was once ranked the eighth-most corrupt official in the world. Given the history of Wyoming Corporate Services' customers, it is shocking that this business model is perfectly legal. Despite America's hard stance on international transparency, the laws within the states remain relaxed. In terms of corporation formation, Wyoming and Nevada have standards lower than Somalia.[104]

Secretaries of State Highlight: R&R Delivery

It sounds like something out of a cartoon, where Tom puts a cardboard 'Cheese 4 Sale' sign in front of a mousetrap, where Jerry will be sure to see it and fall for the scam. In September 2016, scammers took the name of a legitimate Better Business Bureau accredited business, R&R Delivery, and used it to create the appearance of an established business. The fraudsters made a website which advertised a street address and two phone numbers, one with a local area code. When I first found this case, the website was still active. Several months later, when I was writing about it, it had been taken down. When the website was still active, it displayed the BBB's 'Accredited Business' logo, and I have to say, the whole operation looked pretty legitimate.

The fake R&R Delivery contacted individuals in an attempt to hire them, their job being to repackage and reship merchandise. The fraudster asked for their new employees' Social Security number and bank account information, which is pretty standard for a new employer, so they could deposit their checks. Not only did this expose these new 'employees' to identity theft, but it also potentially incriminated them as accomplices to any crime committed.

The scheme works by having the employees forward money, or in this case, reship merchandise that was purchased with stolen credit cards. The employees may not realize the goods or money was illegally acquired. Since the employees are now incriminated, it takes some of the heat off the actual fraudster and adds a middle-man to pin the blame on.

This scam has been featured on a Scam Warners website, where, interestingly enough, the job description for a Forwarding Agent for R&R Delivery was located. This company, which we know is fraudulent, lists both the phone numbers for the real company and the fraudulent one. Scam Warners has placed a large, red banner on it, which tells you this is a fraudulent company. And if you search that same job description, you will find the Scam Warners page for an Albatross Delivery Service featuring the same exact job description—weird typos and all. According to an Advisor from ScamWarners online, "It is very easy to find and use details of legitimate companies—preferably without their own website. By providing a different telephone number, a domain email address, or just a webform, you get anonymity with the patina of respectability."

It is impossible to say whether or not these are related incidents. More than likely, it is a simple template that has been used over and over again, but with very similar results. An employee from the Canadian company Albatross Delivery Services said that within the last six to seven months, they have been getting phone calls from American citizens complaining about them taking their personal information, and not being compensated for work performed. It is the exact same scheme, with the same template, resulting in the same consequences.

Like most fraudulent schemes, this one blends multiple different types together. It uses address mirroring, the technique of having an address that makes it look quite similar to a pre-existing business. It re-routes goods and credit cards through unsuspecting third parties, and it also exposes many of those unsuspecting peoples' personal information. It would not surprise me if the victims of that scam—performing work and not getting paid—ended up having identity theft problems later.

Business identity fraud converges and falls to an unlikely agency: each state's Secretary of State. The Secretary of State is vaguely responsible for state business dealings and created a task force in 2011 within the National Association of Secretaries of State (NASS), to deal with business identity fraud. To do so, NASS produced a white paper in 2012 that really lays it out:

> While retailers will often check with business credit ratings agencies to verify that the information in an application is correct, it can be difficult to immediately detect that a crime has been committed because this information is

> based on the same Secretary of State business records that have been altered by the criminals.[105]

What they mean is that, like Dun & Bradstreet's credit data, the Secretary of State's website relies on self-reported data, which is vulnerable to fraud in every way imaginable.

Businesses have identities just as much as individuals do, and their identity security can be compromised just the same. One of the biggest places of vulnerability? Each state's Secretary of State website, where every business is registered, providing access to tons of identifying information.

Recall the previous section on the Partial Death Master File, specifically where I noted that the Social Security Administration had its hands tied when it came to not publicizing state reported death data on the Partial Death Master File. The overlap is absurd. You may ask, why on earth would Secretaries of State put this readily exploited information on the internet and require little to no security to keep it safe? Well, according to NASS, they have to:

> As long as a document meets certain basic requirements, the Secretary of State's office often has little or no authority to question or reject its contents, to include changes that are made to business filings. Under the Model Business Corporation Act (MBCA), a model law that many states have used as a basis for their specific statutes that regulate business filing and company formation processes, the Secretary of State's corporate filing duties are part of a "ministerial" role with very limited discretion in reviewing the contents of documents.[106] These

> offices also have no authority to control who can view or gain access to state business filings, which are public record.[107]

While it differs from state to state, this 'model' is the basis for law in thirty-two states and the District of Columbia, and is used in part by many others.[108] The most recent revision was done in 2016, and it made no changes to the subchapter that we are referencing. Even more comforting, the 2007 and 2010 model law versions are exactly the same.

So of the four tenets of this model law, which is virtually real law since most states adopted it, a simplified version goes as follows: any documents have to comply with certain standards, and since the Secretary of State's role in all of this is "ministerial," their filing or refusing to file does not:

> 1) affect the validity or invalidity of the document in whole or part; 2) relate to the correctness or incorrectness of information contained in the document; or 3) create a presumption that the document is valid or invalid or that information contained in the document is correct or incorrect.[109]

So if a document is filed and posted on the Secretary of State's website (which, remember, is what a lot of agencies use to double check the validity of the business or person responsible for said business, since it is self-reported), it does not mean that the document is valid or correct. Basically, the fake law that everyone turned into a real law relieves the Secretary of State's office from any form of responsibility whatsoever. Compromising these 'proofs of rite,' such as an address or name listed on the Secretary of State's website, is exactly how business identity thieves get in.

Now, let us examine part A of the Model Business Corporation Act: if it satisfies the requirements of section 1.20, it shall be filed. Have you ever watched *Monty Python and the Holy Grail*? "First shalt thou take out the Holy Pin, then shalt thou count to three, no more, no less. Three shall be the number thou shalt count, and the number of the counting shall be three."[110] That is exactly how this reads. My summary is as follows:

a) Documents need to meet these requirements. Redundant, but go on.
b) Documents must contain the information required by this Act. I directly quoted that.
c) Documents need to be reproducible, and in English.
d) Documents must be signed by the individual in charge of the business and the person executing the document, but does not need to contain a seal.
e) If you use facts, cite your sources, don't make stuff up.

After outlining these sparse, and fairly unhelpful points, the MBCA discusses what exactly is and is not a fact. While I initially thought it was unnecessary, given our recent history and the distribution of 'alternative facts,' I am glad they outline what does and does not count as a 'fact.' Documents must be factual (but they will not verify they are), reproducible, monolingual, and signed.

Those are the only requirements for a non-valid, non-verified, non-correct document to be filed with the Secretary of State. I always wondered at the title, but this all explains it. The Secretaries simply files things away, and do not necessarily know anything about what they are filing. They are the stewards of information, and now, they are trying to retroactively learn about what they are filing, to prevent the fraud that is already rife within their keeping.

Out of curiosity, I started looking up businesses owned by friends on the Wyoming Secretary of State's website. I checked on one friend's business, and online, publicly, where anyone with an internet connection could get it, on a government website no less, I found both owners' names, their address (home and business), phone number, email, filing ID, tax standing, and a load of PDFs including their initial filing and annual reports. Those PDFs included a slew of information all on their own—and this was just a recently started business. Older businesses have much more on file. It also seems that the process to change who your registered agent is (the person responsible to the state for that particular business) is pretty simple. The form was two pages long, and not what I would call detailed.

I looked up a business I knew was no longer active. All the same information, including filing ID and principal agent, was still listed. Then I clicked 'reinstate.' The process for reinstating a non-active business requires no information in addition to what is publicly available on the site. For all intents and purposes, you can reinstate a dissolved business under your name, or someone else's. The website gives you thirty minutes, and there is a video tutorial to teach you how to do it. All you need is the filing ID (publicly available on the business's page), the new address information, a full name to identify as the business's 'registered agent,' payment of the filing fees, and an electronic signature on the document.

This is identity theft of a dead business—the same exact tactic that occurs all the time with individual identity fraud.

With individuals, we refer to it as 'ghosting' or 'zombies'—essentially bringing the dead back to life. So what happens when you revive a dead business? The easiest way to revive a dead business is through the Secretary of State's website, specifically with Limited Liability Corporations (LLCs).

LLCs are a way of keeping an individual, like a taco truck owner, and the business of owning a taco truck, separate. That way, if you, the taco truck owner and taco chef, make a taco that gives someone food poisoning, the poisoned individual cannot sue you and take all your stuff. They can only sue the taco truck and its assets.

LLCs have other benefits as well, mainly that they offer pass-through taxation. Most corporations get taxed twice, once when they make profits, and a second time when the dividends (all the money they make off those profits) gets passed through to investors. LLCs, on the other hand, do not pay dividends to investors, so the owners only pay taxes once. LLC members also do not have to be United States citizens or permanent residents. They require much less paperwork because there are fewer regulations surrounding them.

All these components make LLCs a great way to start a company, especially for someone starting a small business like a taco truck. However, the lax regulations also make it incredibly easy to start fraudulent endeavors, or to compromise existing, law-abiding endeavors. LLCs are the jumping point for a lot of business identity fraud.

Unfortunately, LLCs are especially vulnerable to this kind of fraud. A fraudster can add themselves as an officer to the Secretary of State's webpage. Once your name is listed there, obtaining credit cards and/or loans, and then cashing them out and disappearing is not difficult to do.

This same method applies to real estate and property fraud: finding a million dollar condo and the LLC that is in control of it, adding yourself to the Secretary of State's website, and then going out to get a loan. According to Richard Petrovich, a lawyer who focuses on complex business litigation, Florida has a particularly serious problem with revived businesses. Florida has the highest elderly population in the country, and[111] one of the highest individual identity theft rates as well. In Florida, and in many other states as well, Mr. Petrovich tells us that, "All there is is a warning if you use information falsely. There is no identifying information. At the very least, you should have some kind of a pin number to allow you access, but that does not exist."

Single-asset LLCs, typically formed to hold property as the only asset, are extremely popular in South Florida, which just so happens to be the second most popular place to commit mortgage fraud. All the information you would need to transfer ownership is on the "About" page of any business's website. Identity thieves can search through public records and find real estate without any mortgage attached to it. If the property has equity that can be stripped, fraudsters can apply for loans secured by mortgages on the property.

For a single-asset LLC, you simply add yourself to a position of authority for the LLC, and that change will be visible to anyone who is verifying you and your business through the Secretary of State's website. Apparently, no one will actually verify your relationship to the LLC, as long as you are listed on the Secretary of State's website—something that sounds very familiar to verification issues concerning documents and identity.

That is part of the reason why this is so fascinating. Across all forms of fraud, there are always the same weaknesses: an ease of disguising your identity and a lack of verification on the other end.

Rather than looking at business or individual identity fraud in a linear fashion, try to understand it like a spider's web. Every segment is somehow connected to every other segment, but on opposite sides of the web are two boxes equidistant from the center: One labeled "Individual" and one labeled "Business." Insinuating yourself within an LLC to take out a loan, on the surface, sounds very different than stealing your neighbor's identity, but invariably, the principles that allow them to occur are the same.

Trust in Me

Businesses rely on trust. You have probably seen the headlines within the last five years mentioning giant corporations and data breaches. If your customers do not trust a business, they will not work with or buy from them. Since most markets are competitive enough, another company will swoop in and take advantage of the loss.

Government agencies do not have the same problem, or at least not to the same degree. Plenty of people distrust the government, but there is only one federal government, and the lack of competition correlates to a lack of initiative in fixing their faults. While trust from their constituents is important, it does not make or break certain agencies the same way it does with businesses.

The critical component of trust breeds an interesting conundrum: often, businesses try to cover up when business identity fraud (or data breaches, because remember, they are different) happen to them. They do this to protect their own assets, understandably, but it does make it difficult to report on the statistics of business identity fraud, since no one wants to own up to it.

Many businesses would choose not to report these kinds of problems, rather than risk the bad press and possible consequences that come along with it. Studies are also unlikely to further separate out business identity fraud schemes from the ones that are solely perpetrated by misuses of the Secretary of States' data, making the parsing of data less specific.

Many businesses, like discouraged individuals dealing with identity theft, come to view fraud as something they are powerless against. A Kroll white paper likened it to the challenge faced by IT professionals, who have to protect network security while ensuring that it runs smoothly and does not slow network traffic. It is the balancing act faced by many people: providing quality and safety, while at the same time maintaining speed and efficiency.

It is a conundrum I had to remind myself of while stranded in the grocery store on a July day with a cart full of perishables, on the phone with my credit card company who thought my trip to the same grocery store I go to all the time was suddenly fraudulent. The chip in my card had stopped working, which halted my purchase, and I had to speak with Jim at the credit card company to verify my identity. Chip technology does great work preventing fraud, but when it inconveniences us and is melting my Bagel Bites, some of us might lose our cool.

Having fraud preventions that interrupt daily operations might drive some customers away, but it is far better than facing the repercussions of losing control of customers' data. After the Bagel Bite incident of 2017, I started going to the smaller, local grocery store for the sole reason that they did not have chip readers, and I could just swipe my card and get on with my day. I was slightly inconvenienced by fraud protections, put in place to prevent me from having my identity stolen (which is a way bigger problem than having to call my credit card company and talk with my friend Jim) and so I inadvertently decided to change grocery stores. Reflecting on the series of events, I am a little appalled at my laziness.

This is exactly what businesses are facing—except instead of melty Bagel Bites, it could be huge amounts of money, or even an entire business. That same Kroll white paper noted that:

> Many businesses have come to believe that the cost of fraud, though high, is not as high as the cost of friction in the sales process. For these businesses, fraud is simply an unfortunate but unavoidable cost of doing business.

This puts businesses at an even higher risk than they realize. Ignoring the root cause only makes it worse. Hiding fraud for fear of the consequences can affect the inside operations of a business. If an employee covers up fraud for fear that having missed it will get them fired, it may in turn affect how the boss thinks about fraud risk protections. While hiding it from your customers will, for a short while, keep their trust, in the long run it might put them at risk and lose their business for good.

One company I spoke with had those same reservations, but concluded their customers' safety and security was most important. In 2011, Scott Burnett, co-owner of AAA Termite & Pest Control, opened the latest edition of the yellow pages and realized there were three other AAA Termite businesses listed. It appeared that additional branches had opened and were doing business. Mr. Burnett was proactive. He reached out to local and national news, did television interviews, and even spoke with NPR.[112] That is not to say that Mr. Burnett had his doubts, "Obviously we do not want our customers to know there are fake branches of our business, because we want our customers to call us. You want your client to go, 'Oh, they are on top of this, and if they send someone out, we can trust them.'"

Luckily, those reservations did not stop Mr. Burnett from reaching out. He contacted local law enforcement and was met with continual frustration: "It's against the law to impersonate a licensed official," Mr. Burnett said, after noting that pest control agents are licensed, "but it's a Class C misdemeanor, so it is not that important to them. Besides, no one had even heard of this happening before.[113]

In addition to the three other AAA Termite businesses, 103 fake pest control businesses appeared in the yellow pages. When Mr. Burnett called the phone company to try to get the address of the fake companies, armed with proof that he was the legitimate business and the others were fraudulent, including an attorney and the backing of the National Pest Management Association, the phone company would not budge. Privacy laws protected them.

Mr. Burnett believes nothing came of the scheme because they jumped on it so quickly and went after the imposters so aggressively, but he admits that, with fraud, you never really know what is coming down the pipeline. Mr. Burnett actually had his personal identity stolen previously, which was why he was so adamant about warning his customers and telling them to be very certain about who they were dealing with before giving out a credit card number or other personal information.

The most interesting part of my conversation with Mr. Burnett came right when we were about to hang up. He told me that he had been hesitant to return my phone call, and was lying in wait for me to reveal my fraudulent intentions by asking him about his personal information. "After this," he said, "you're always looking over your shoulder. You're always checking everything."

Mr. Burnett and AAA Termite might have the best answer for fighting business identity theft on a small scale. Maintaining vigilance and spreading awareness might make it more trouble for the fraudsters than it's worth. While no one wants to admit their business was defrauded, being forthcoming about it and protecting your customers like Mr. Burnett did, in the end, is the best thing for everyone involved.

We do have to consider that, unlike with a single individual's identity, businesses cannot be content with focusing on recovery if they have been defrauded, or prevention if they think they may be in the future. Individuals have the luxury of not having to make a profit throughout the process of being defrauded the way businesses do. That is not to say it does not disrupt an individual's life—certainly it does. Having your identity stolen causes a massive loss of time and an increase of stress, but at the same time, it often does not impact an individual the way it impacts a business.

The bigger problem is silence. If it is a secret within businesses, we cannot systematically address it. We cannot collect adequate statistics that can be used to fight fraud, advocate for better protective systems, or even find the issues within the current systems. Better, more honest data will help address these problems at the start, rather than hiding them.

The Case of Adam and Corporation X

We spoke with an executive of a mid-sized corporation about this very problem. We will call him Adam, but neither Larry nor

I even know his real name, or the company he works for. We will call the company Corporation X. We originally spoke with a friend of Adam's who told us his story, but after insisting we would keep him anonymous, he finally agreed to speak with us himself.

Adam is an executive at Corporation X, which does roughly $500 million in business a year. One day, Adam received a letter from the IRS looking to verify that his company had applied for a quick refund, which is a refund given for overpaid estimated taxes. The letter stated that Corporation X had 20 days to respond. Adam was shocked and contacted the IRS that same day. After insisting they did not file for the refund, the representative asked if Adam had ever filed the company's power of attorney before. He said yes, but not recently. When Adam asked who the power of attorney was listed as, the representative told him they were not authorized to release the name of the power of attorney to him.

Adam asked whether or not the money had been released, and was assured it had not been, and to sit tight. Adam made the first call on a Wednesday. On Friday, he heard from a different representative. This new representative asked him similar questions: "If you filed for power of attorney, who would it be? What bank would the refund be sent to?" None of his answers matched what the IRS had on file. Finally, the representative told Adam that the money had been released, in the amount of $3 million.

Luckily for Adam, his company, and the IRS, Adam acted fast enough that the IRS was able to take the money back out of the account they had wrongly deposited it into. Had Adam waited much longer to call, the money would most likely have been transferred out, and irrecoverable. Still, many questions remained. The form the fraudster used to apply for the refund was form 4466, 'Corporation Application for Quick Refund of Overpayment of Estimated Tax.' This form allows a company that has overpaid estimated taxes for that quarter or year to get a quick refund on the overage.

Form **4466** (Rev. October 2016)
Department of the Treasury Internal Revenue Service

Corporation Application for Quick Refund of Overpayment of Estimated Tax

► **Information about Form 4466 and its instructions is available at *www.irs.gov/form4466*.**

OMB No. 1545-0123

For calendar year 20 or tax year beginning , 20 , and ending , 20

Name	**Employer identification number**
Number, street, and room or suite no. (If a P.O. box, see instructions.)	Telephone number (optional)
City or town, state, and ZIP code	

Check type of return to be filed (see instructions):
☐ Form 1120 ☐ Form 1120-C ☐ Form 1120-F ☐ Form 1120-L ☐ Form 1120-PC ☐ Other ►

1	Estimated income tax paid during the tax year			1	
2	Overpayment of income tax from prior year credited to this year's estimated tax			2	
3	Total. Add lines 1 and 2			3	
4	Enter total tax from the appropriate line of your tax return. See instructions		4		
5a	Personal holding company tax, if any, included on line 4	5a			
b	Estimated refundable tax credit for federal tax on fuels	5b			
6	Total. Add lines 5a and 5b		6		
7	Expected income tax liability for the tax year. Subtract line 6 from line 4			7	
8	**Overpayment of estimated tax.** Subtract line 7 from line 3. If this amount is at least 10% of line 7 **and** at least $500, the corporation is eligible for a quick refund. Otherwise, do not file this form. See instructions			8	

Record of Estimated Tax Deposits

Date of deposit	Amount	Date of deposit	Amount

Sign Here

Under penalties of perjury, I declare that I have examined this application, including any accompanying schedules and statements, and to the best of my knowledge and belief, it is true, correct, and complete.

Signature | Date | Title

General Instructions

Section references are to the Internal Revenue Code.

Who May File

Any corporation that overpaid its estimated tax for the tax year may apply for a quick refund if the overpayment is:

- At least 10% of the expected tax liability, and
- At least $500.

The overpayment is the excess of the estimated income tax the corporation paid during the tax year over the final income tax liability expected for the tax year, at the time this application is filed.

If members of an affiliated group paid their estimated income tax on a consolidated basis or expect to file a consolidated return for the tax year, only the common parent corporation may file Form 4466. If members of the group paid estimated income tax separately, the member who claims the overpayment must file Form 4466.

Note: Form 4466 is not considered a claim for credit or refund.

Cat. No. 12836A Form **4466** (Rev. 10-2016)

Form 4466 can be used to file a quick refund.

The fraudster first filed for power of attorney. Most of the information necessary to do this is publicly available. In addition, the only personal information the IRS asked of the fraudster was his name, title, and address of the company: all things he could easily obtain online. The fraudster used the stolen identity of a pre-existing power of attorney from New York: a seamless combination of individual identity and business identity fraud sewn together to look realistic. They even went so far as to set up a phone number with the correct area code for Adam and Corporation X, to go directly to a voicemail with a very official sounding, yet perfectly vague message: "You have reached the CFO for Corporation X." If you are an IRS employee calling to verify someone's identity, that would probably be enough information for you.

The IRS investigators managed to learn that the scam originated in London, meaning there was no way they would catch the criminal, despite recovering the money. Even though the scam came from the United Kingdom, the bank account was American. The most likely scenario was that the British scammer advertised a job requiring an American to set up a bank account, which they would later use to house the fraudulent refund. This is reminiscent of the R&R Delivery scam: in this case, whoever set up the bank account might be unknowingly incriminated.

The timeline went a little like this: the fraudster filed for power of attorney using fraudulent identity information. Then, ten days after filing the power of attorney form, they submitted Form 4466 for a Quick Refund, which does not contain a whole ton of identifying information. Their way of verifying the power of attorney was to send Adam at Corporation X the letter, carbon copying the power of attorney, who was the fraudster undercover as an unsuspecting man in New York.

The other verification they have in place is just as pointless, given the fraud at play. They wanted the fraudster to verify the amount of the refund and the bank to which it was going. Of course they could verify it, since they were the ones that filed it. Adam could not even verify it, which makes this security measure more theft-friendly than safe.

When I asked Adam how this has affected his company, he chuckled. "It was scary, and now I'm much more appreciative of the layers of security. We are going to put in cyber insurance, which we did not have before. We are much more mindful of what we send in emails now." He did also point out that cyber insurance would not have helped them even if they had it.

Since it was not the company's blunder, but that of the IRS, it would not be covered by any insurance the company had. Instead, it would be covered by the insurance of the government. Had they not recovered the funds, the IRS would have had to pay the company back for their mistake. Taxpayers, in the end, would be paying the $3 million in losses. Not only would taxpayers foot the bill, but they probably wouldn't even know they were. The IRS isn't exactly known for their transparency when they increase the burden on taxpayers.

As long as we are comparing business and government operations, let us note that had Adam's company been found at fault, they would have had to foot the bill. As with any kind of decision, businesses have to balance cost analysis. Governments, while also responsible for allocating funds, will not be held accountable in the same way as an employee or a business. The IRS has a sordid history with problematic programs and problematic solutions.

An unfathomably dense TIGTA report from 2017 identified some of the problems the IRS has had with improper payments. The Treasury scores the IRS's different programs on a risk scale. To be categorized as a high risk program, one must "exceed both 1.5 percent of program outlays and $10 million of all program or activity payments made during the fiscal year reported or exceed $100 million at any percent of program outlays." To even be considered high risk, you essentially have to throw away $10 million.

One of these high risk programs is the Earned Income Tax Credit program. The IRS estimates that 24 percent, or $16.8 billion of Earned Income Tax Credit payments are improper—and that is only one IRS program. Even then, TIGTA believes that the IRS's method for conducting risk assessments is flawed. The IRS evaluated their own programs and rated several as low risk. When TIGTA evaluated those programs in 2015, using the IRS's own data, they found that one of those supposedly low risk programs had 24.2 percent of improper payments, totaling $5.7 billion.

In 2016, that same program had an increase to 25.2 percent, or $7.2 billion improper payments, and this is apparently low risk in the eyes of the IRS.[114] Not only does the IRS have a significant chunk of improper payments, but their way of reporting them is deeply flawed. The first step to recovery is recognizing the problem, but the IRS is addicted to deflection.

The Chief Financial Officer for the IRS cited the agency's limited resources and ability to correct and audit returns, lack of tools for verifying data, and that the difficulties in administering IRS programs "frequently stem from how Congress structured them."[115] Tackling improper payments on an enormous scale is no easy task, but low-balling their high risk programs makes it look especially fishy. Adam's fear of repercussions and hesitancy to speak with us was actually far more interesting than the story itself. It fully proved our point: the corporate culture of keeping silent is causing a bigger problem. By keeping secrets about business identity fraud to protect one company's reputation, all corporations suffer from lack of awareness.

If no one reports their numbers, there can be no analysis from those statistics, meaning employees, employers, lobbyists, politicians, and anyone else with a vested interest has no accurate numbers to point to when trying to prove that business identity theft is a real problem.

One white paper, produced by the Identity Theft Research Center, explores the idea of problematic statistics through property crime rates. The *Crime in the United States* report, which the media uses to cite crime statistics, does not include property taken by fraud in its category of property theft. In this vein, fraud-related crimes may be increasing, but since they are not included in the category of property crimes, the general statistic goes down.[116] Over 16.6 million people were victims of identity theft in 2012, incurring losses of $24.7 billion, meaning property crimes should have spiked that year. Instead, FBI reports show a decline of 14.1 percent of property crimes at that time.[117]

The statistics used to measure identity theft in that instance only focused on financial identity theft. It left out all the other types, such as medical, government, and tax. The statistics surrounding fraud are flawed. Companies and agencies do not accurately report their numbers, subjecting these statistics to scrutiny, and boxing them into incorrect categories. We like to think statistics are hard numbers, inarguable, but this is false.

This report spurs headlines like "What Caused the Great Crime Decline in the U.S.?" and era nicknames like "the Great American Crime Decline."[118] While yes, violent crime rates may have dropped, we can also see an increase in what was previously known as 'white collar crime,' like identity fraud.[119] There has long been a connection between criminals involved in drug operations that have since switched over to organized identity theft. The risks are smaller and the returns are bigger.[120] In 2015, the state accused the Long Beach-based Crip gang of identity theft and tax return fraud to the tune of $14 million.[121] According to one Miami detective who has worked extensively with South Florida gangs, conducting a traffic stop and finding tax returns instead of drugs has become a commonplace occurrence. In 2015, the Van Dyke Money Gang made more than $1.5 million with a Western Union money order scheme.[122] With so many of the larger crime syndicates turning to less violent means, it makes sense that violent crimes and property crimes are diminishing while fraud rates rise. Having statistics that do not accurately reflect identity-related crimes can impact whether individuals and businesses decide to invest in identity protections, who might rather wait until a scary headline spurs them to action. If a business owner does not have access to real information about fraud, how can they make an informed decision about protecting themselves from it?

Kroll's 2015-2016 Global Fraud Report found that "fraud has continued to increase, with three quarters (75 percent) of companies reporting they have fallen victim to a fraud incident within the past year."[123] Business identity fraud has remained a swept-under-the-rug type of fraud. Because businesses are often either unaware of an occurrence, or because they don't want to ruin their consumer trust. No one wants to be the bearer of bad news.

The same Kroll report notes this trend: "No survey can ever measure unproved and undiscovered fraud, probably the largest categories of all, making loss statistics questionable." Having no trustworthy statistics makes it difficult for businesses to care about fraud. Why carve out a budget for a threat that may not even exist?

6

Teamwork Makes the Dream Work

Fraud prevention is managed and contained within industry silos—and often only reactively, responding to the latest fraud trends and schemes. An industry-only view is equivalent to staring at one's navel.
— Vikram Dhawan and Carlos Garcia-Pavia, "What Financial Services and Insurers Can Teach Each Other," 2015

I remember teachers telling me and my classmates that group work is essential, that learning how to work together is a life skill. I, the overachieving nerd that I was, felt like group projects were simply a way of raising the lazy bumpkins' grades by pairing them with students like myself. I hated group projects. Unfortunately, Ms. Rose may have been right, teamwork is a life skill, but I am not the only one who has struggled with it.

Individuals and businesses are affected by identity fraud, but so are local, state, and federal governments, which means that taxpayers are affected, too. Taxpayers meaning the individuals who own, support, and patronize businesses. So, as you can see, no one is safe from the vicious cycle of consumption and fraud dependency. And everyone, it seems, has trouble working together. Fraud is continually broken down into smaller and smaller categories: tax refund fraud, identity theft, hacking. We fail to recognize the larger connections at play. Dividing our knowledge into individual industry silos hurts more than it helps. In this group project, it is not "divide and conquer," but "combine and prosper."

Tax refund fraud, for example, is connected to every group in that list. As we know, an individual can have their identity stolen. A fraudster can then use that stolen identity to submit a fraudulent tax return using a business's EIN, which then gets that business in trouble and loses the government money. The more we break down fraud, the further away we get from collaborative learning and sharing.

Business and government share a similar set of problems, many of which are inter-connected. What do business and government each do well in the fight against fraud, and what can they learn from each other? What strategies have both come up with? Even between industries like insurance and health care, there's a lot to be learned from each other's successes and failures.

The best part about sharing methods to deter and stop fraud is that if one sector succeeds, it helps everyone succeed. It's not like a group project where I do all the work and stupid Jeremy also gets an A (though perhaps individuals working in each sector may occasionally feel like they are pulling more than their share of the weight), but instead it's like many players working together to defeat a greater evil. Ms. Rose was right. If the ship sinks, every member of the crew goes with it.

Preventing fraud, in all of its ugly mutations, is better for everyone, in every industry, at every level. If we can prevent fraud from occurring, the government has more money and a greater ability to fund programs and give out rebates that help individuals succeed. If individuals succeed, they can better turn around and support and run businesses. It is a happy, fraudless circle of life. That might seem like a simplistic overview, and it is. Obviously, there are a million different components to what fraud does to certain industries, but the overall point is that fraud is bad for everybody except the fraudsters.

A lack of communication is at the base of all industry problems, but businesses in particular have trouble talking about their problems. As we discussed, the stakes for businesses are far higher than they are for government agencies. Of course, government agencies and institutions are held accountable by their constituents, but there is only one government. For a business, the ever-present fear of losing money, or getting chased out by a competitor, is more real. Because of the competitive nature of our open market, the stakes are far higher for businesses.

It is uncomfortable to feel scrutinized as an employee, especially when it is unwarranted. Unfortunately though, employees are a serious component of identity fraud. In 2007, an employee at a Jacksonville, Florida branch of Fidelity National Information Services stole 2.3 million consumer records, containing credit card, bank, and other personal information. The consumers affected by this scam received solicitations from companies that bought the data, but at the time, the president of the company said, "We believe that is the extent of any damage to the public."[124] Knowing that identities accrue more value over time, it is likely that Fidelity purchased identity protections for the individuals whose identities were compromised for two years, and then discontinued it. Two years is the benchmark for companies following up—but those identities aren't safe after two years, and unless the individuals start purchasing it, they are still vulnerable.

The employee who sold the information was a senior level database administrator who had worked at the company for seven years. These kinds of stories are disarming. They disrupt our idea of loyalty and trust. Obviously there was no probable cause to believe the employee would do this kind of thing, otherwise he would have been fired ages ago.

Similar to 'familiar fraud,' business insider fraud is more insidious than one would imagine. Forty percent of United States respondents to the Kroll Global Fraud report said that a senior or middle manager had been a major player in at least one such crime.[125]

This is something that government agencies deal with as well. We have seen several examples of DMV employees collecting money on the side to help produce fraudulent documents. In one case, a DMV worker charged people $2,500 a piece to create fraudulent licenses, netting roughly $34,000 from her side-hustle.[126]

It is a human condition we all struggle with, in or out of the office. How can you ever really know if you can have faith in someone? There are never any guarantees with trust, you just have to, well, trust. Putting the great human struggle aside, there are protections businesses and government agencies can put into place to help deal with their trust issues.

The Five C's

Dun & Bradstreet published a white paper titled 'Mitigating Business Fraud,' in which they discuss a model plan that they believe will help businesses fight identity fraud. "The Five 'C's of Fraud Prevention" is a strategy developed for government agencies that was initially adapted from commercial best practices for credit management. It also acts to "provide companies with a strategic framework to guide their efforts to incorporate verification and authentication activities within existing organizational processes and information technology systems." What's good for the goose is good for the gander.

"The Five 'C's of Fraud Prevention" is a good start for all agencies, government and business alike. They include Confirmation, Condition, Consistency, Character, and Continuity. As far as alliterative guidelines go, these are pretty good, but they are far from bulletproof. Many industries think they already follow these tenets, without realizing how many holes the protections they have in place leave.

1. Confirmation

"The Five 'C's" tells us that "a quick check of databases will reveal if a particular business is registered to operate in the state or has a valid address." This is a good start, but think about the frauds we have discussed. I could confirm a business on the Secretary of State's website, but that data is self-reported, and thus unreliable.

The intersection of government and business is a crucial component to Confirmation, since now a system operated by the government (the Secretary of State system for listing businesses online as public record) influences what businesses are listed, how they are listed, and who they are listed under. We can try our darndest to vet businesses before dealing with them, but if the way we confirm details about a business is vulnerable to fraud, there is not much else we can do.

The first ‘C’ also suggests checking to see if the business has a valid address. With address mirroring, you can check all you want, and you will still see a valid address. It might just not be the right valid address. Fraud is terribly interesting in this respect. By finding the holes in the net that help individuals feel at ease, by influencing something like the Secretary of State’s website, fraudsters are messing with the very thing that is supposed to make us feel secure. “It’s on the Secretary of State’s website, of course it is legit,” we tell ourselves. If it ends in ‘.gov,’ it’s the real deal right? Maybe not.

The problem with Confirmation is that in these intersectional regions where multiple industries and agencies have to rely on each other, everyone has to be on the same page. It appears that businesses have a higher standard of Confirmation in some ways than the government, so if the government cannot get on their level, businesses start to lose out, which eventually affects their ability to trust in the agencies they deal with.

2. Condition

The second ’C’ is Condition, and shares many of the same problematic qualities as Confirmation. Condition refers to whether a business’s status is active or not. Does it have a functional telephone number, website, or e-mail address?

These investigative questions will filter out a good portion of fraudulent enterprises, but there are plenty of examples where they would not have. Shell and shelf companies do their best to give the appearance of being a legitimate business. They have active phone numbers and email addresses. That is all part of the scam. Even if a business is revived as a zombie business, their status will appear as ‘active.’

3. Consistency

The third 'C' is much more solid than the first two. Consistency is key, and looking at a business's track record is telling. Traditional 'bust-out' scams involve managing a credit file—which can vary from the individual's own name, a stolen identity, or a synthetic identity—for long enough to establish a satisfactory credit limit. Once they have that, they will max out the account and disappear.[127]

How long the business has been in operation, whether it has reviews, and how long those reviews have been around provide good information. Perhaps the reviews are fraudulent, but many websites that allow reviews keep a pretty tight lid on allowing reviews that seem suspicious. Their reputation as a trustworthy reviewing entity is at risk, too. An Experian report points out that, "bust-out fraudsters have almost five times as many inquiries as the good accounts."[128] The report also notes that bust-outs are not typically associated with property, like mortgages, home improvement loans, etc.

The Dun & Bradstreet Five 'C's list gives a perfect example: "Does the company president also run 15 other businesses registered at the same office? This is a definite warning sign that something is awry." The example from the previous chapter about the Cayman Islands of Wyoming, Cheyenne, checks that box. The gentleman in charge of Wyoming Corporate Services is listed as either a director, president, or principal for at least 41 out of the 700 companies the incorporating services lists for sale. Who has time to be in charge of 41 (really 42, since he runs the incorporating service) businesses?[129] Which takes us to the other thing the list mentions: if it sounds too good to be true, it probably is.

4. Character

Another more reliable component on the list is Character. There are ways to fraud your character, but it is far trickier than making a real phone number and email address. Dun & Bradstreet recommends conducting criminal and background checks. We agree, and also recommend pulling a credit report. While the Better Business Bureau provides an invaluable service and is a great place to start an investigation, there are ways around it. The fraudulent R&R Delivery company, for instance, had a Better Business Bureau icon on their webpage. Most likely it was a simple cut and paste job, but the impression it gives is a powerful one. Seeing the Better Business Bureau label on a page, much like seeing a business on the Secretary of State's website, can lend an undeserved air of legitimacy.

In the case of R&R Delivery, there were two Better Business Bureau pages: the page for the actual business, and a second page for the fraudulent business. If you are not investigating this as thoroughly as I have been, and you are busy at work, and you are thinking about what to make for dinner, and whether you remembered to turn off your coffee maker this morning, you might miss things. We also have to remember that fraud accounts for human error. We often forget that all the systems we deal with are made by people, fallible people. The customer service rep? Yeah, she's a person. We miss things, we forget things. We can be thoughtless. Fraud anticipates this, and it takes full advantage when we make our human mistakes.

5. Continuity

The fifth and final 'C' of fraud prevention is Continuity. Dun & Bradstreet ask, "Has the operation's status changed and is it posing new risks? Just because you've signed the contracts, it doesn't mean your job is done. Implement a system that continually monitors contacts for events that may indicate potential trouble." This, along with Character and Consistency, is the strongest check you can have. Human fallibility, remember?

You can make it work to your advantage as well. Humans are creatures of habit, and since companies are human endeavors, they can make mistakes too. If a company you have always dealt with suddenly seems different, maybe they ordered a batch of something out of character, like rather than fifty reams of paper they ordered five hundred. Maybe credit cards are being taken out in a way they have not normally. Patterns are important to pay attention to, so if something seems out of place, check on it.

Dun & Bradstreet notes their list is not perfect; no anti-fraud list ever will be. They note that this list is a "great foundation" (which we fully agree with) but that "companies also must adopt clear policies for applying their fraud prevention rules and processes." No policies will work if staff does not know how to or does not bother to implement them well. Even then, there is always some risk. Mitigating that risk, however, is totally worth it.

These 'C's are great concepts to keep in mind when vetting a business, but one of the most important things to remember is that fraud is designed to deceive, and if a fraudster can influence a verification system in any way, it only makes the fraud stronger. Fraud will exist in unexpected places. Staying vigilant, listening to gut reactions, and questioning the norms may serve companies and individuals well.

Before[130] we got so focused on these five 'C's, I quoted Dun & Bradstreet in saying that these strategies were "developed for government agencies," and also that they "can provide companies with a strategic framework." These strategies work for both government and businesses. Rather than looking at the two as such different entities, we need to look at them together, especially when you realize how much one depends upon the other. Sharing successes (and failures) across the board will help everyone do a better job of preventing fraud.

The Pay and Chase Model

The 'pay and chase' model of fighting fraud is a perfect example of a system that does not work. It does not work for business, and it does not work for government. It does, however, work wonderfully for fraudsters. Pay and chase is what happens when an entity gives out goods and services first, and expects to be paid after the fact, only the payment never comes. If someone uses fraudulent information, they cannot be tracked down to pay for something, hence the chase.

Pay and chase exists in every industry and every agency.[131] Federal and state benefits programs also have to deal with pay and chase, as do any businesses that allow their customers to pay invoices after items have been received. This is how bigger businesses can get in trouble—if a fraudster assumes the identity of a "trustworthy" entity, like a large business that always pays its bills, the trust is built into the reputation.

Pay and chase is an overarching model that is equally as ineffective in every industry it exists within. The GAO, in as sassy a footnote as a government document will allow, defines pay and chase as the "labor-intensive and time-consuming practice of trying to recover overpayments once they have already been made rather than preventing improper payments in the first place."[132]

The key part of that definition is that these payments never should have been made in the first place. If payments were made before goods/services/benefits were given, this entire process would not have to occur. Chasing down improper payments and retrieving goods that were mistakenly given out is never easy. I used to wait tables, and reclaiming money that is already spent is like recovering a burger that's halfway down a hungry man's throat. You're just not getting the burger back. It's just as futile as recovering benefits and goods that were fraudulently obtained and already half-digested.

When agencies attempt to recover lost money with the pay and chase method, roughly 17 percent ever gets recovered. This model[133] is impractical and requires superfluous time and money when a front-end solution would save on both.

During the Obama administration, White House officials declared that they were abandoning the old model of pay and chase, and directed agencies to strengthen controls to prevent fraud before it happens,[134] but in 2016, the Centers for Medicare & Medicaid Services were chastised for their use of pay and chase methods when it came to healthcare fraud.[135]

According to Beryl Davis, a Director of Financial Management and Assurance at the GAO, "The ideal situation is to not have to chase the money. It is far better to have internal controls in place that will prevent improper payments from occurring in the first place." Some of the strategies put in place to help prevent pay and chase tactics are being thwarted. The Do Not Pay Initiative, something we discussed when looking at what agencies have access to the SSA's Full Death Master File, is missing three crucial databases that would give a complete picture of who not to pay when it comes to government benefits.

Every agency still has some aspects of pay and chase,

though there have been efforts to reduce it. Often it comes down to whether or not systems operations have been fully implemented to avoid pay and chase, whether employees are correctly trained on the new systems, and if each agency has the time, man-power, and finances to implement them. It is rare for an agency to have each of these factors in line.

Inter-Agency Communication

An article included in the 2015 Journal of Insurance Fraud chronicles the importance of inter-agency communication. It takes a close look at the strategies employed by different agencies and how they could positively impact others. It discusses the use of automated data within the financial sector, and how it not only improves the user experience for customers, but it also helps fight fraud. In terms of implementing automated technology, banks' hands have been pushed by regulations and legal pressure.

According to the article, automated data is no small feat: "It's daunting. Banks must verify and authenticate identities instantly, and fully screen people and businesses before making critical lending decisions." It is a huge analytics task, but since banks have had the pressure to do so, they have accepted it and done so admirably. I can check my bank account on my phone in seconds. When the great Bagel Bite incident of 2017 went down, sure, it was inconvenient, but it was far easier to deal with than having my credit information compromised. Analytics are a strong front-end strategy to prevent fraud, and according to this journal:

> a bank might notify a customer when someone in China used her credit card though the customer used it five minutes earlier in her hometown of Sacramento. Data-driven analysis greatly reduces reliance on human intervention… Mitigating fraud thus becomes far more proactive, and effective.[136]

The financial sector has been so pressured to adapt and move to automated data, but other industries do not have the same level of expectations. As the article points out, the insurance industry still relies on paper copies. Can you imagine, *faxing*? In their words, "A suspicious claim…triggers a manual review by an adjuster and investigator… Less automation means more subjectivity, more time and financial resources, less scalability and lower customer satisfaction."

When banks started running into verification problems, such as not knowing if a person logging into an account was actually the account holder, they started implementing Knowledge-Based Authentication (KBA) questions. We discussed KBA with regard to identifying documents, particularly when Larry ordered his mother's birth certificate online. In theory, KBA asks questions to which only you would know the answers. Questions like, "Which of the following addresses have you lived at?"

KBA questions are excellent, though often, they are mined from publicly held data, which means they are also available to identity thieves. We suggest making these questions more personal, something that cannot be found online. Some institutions allow the customer to select the question and type in their own answer when they sign up for an online account.

Something like, "Who was your third grade teacher?" is more secure even than "What elementary school did you attend?"

One of those questions can probably be answered by knowing what town the individual grew up in, which can often be answered by a Facebook page. Personalizing KBA questions adds another layer of security to the process without adding enormous cost to implement.

Insurers, maybe while being a bit behind on automation, are much better at sharing data. The article highlights efforts made by the insurance industry to share data in order to fight fraud on a larger level. The Coalition against Insurance Fraud and the National Insurance Crime Bureau make a point to compare information to help everyone combat fraud.

The article's whole point is that fraudsters are not picky, and they do not limit their activities to the industry silos we assign them. Fraud will go where the opportunities arise. Most fraudsters cross over from industry to industry. If someone opens credit cards, odds are good they might also sign up for benefits, or file for a fraudulent loan. Fighting fraud does not mean sticking to your particular group, but looking outside of your people and comparing notes to the agencies around you.[137]

Justin's Identity Fight

I am lucky enough to not have had personal experience seeing the financial sector's prowess in action, but a friend of mine has. In 2015, Justin had his identity stolen, but he would not realize it until two years later. Unsurprisingly, since we know identities gain value over time, the culprit waited two years almost exactly before acting on the information they procured.

In January 2017, Justin checked his Credit Karma account, a website that allows you to keep tabs on your credit. Justin had received an email from Credit Karma informing him that there had been an unusual amount of activity on his account. When he checked, he saw a flurry of hard inquiries from stores from which he had supposedly applied for credit cards. Justin, who would not step foot in a department store to save his life, knew something was wrong.

The same day he received the notice from Credit Karma, he also received an email from Wells Fargo, his primary bank, asking if he had made a change to his address. This tactic should sound familiar—it is exactly what happened to Adam and Company X. By altering the contact information, the fraudster automatically becomes the point person. He called Wells Fargo and told them no, that he had not had a change of address, and also told them about what he was seeing on Credit Karma. Wells Fargo immediately froze his accounts, cancelled his cards, and overnighted him a temporary card.

Justin then had to go in person to a branch with multiple forms of identification and proof of residency for other addresses he had lived at, answering KBA questions like, "Have you lived on X street?" and "Have you had a utility bill in X city?" I interrupted Justin to explain how KBA questions are mined from public data, meaning the fraudster could have answered the same questions. He told me that the fraudster had in fact known the answers to those questions, because they are the same KBA questions they were asked when changing his address on the account.

Justin realized at that moment, when he knew he had all his previous address information, that his identity had not been stolen recently, but two years before. He was in Las

Vegas, moving at the time, and his car had been broken into. Because he was moving, he had all his personal, banking, and tax documents, his passport and Social Security number, and wallet, in the car. While the thieves left other valuable items, they found what was truly valuable and ran with it.

After positively identifying Justin in person, Wells Fargo had him create a verbal password, as in, a password that is not written anywhere. Whenever he logs into his account to make changes, he has to say his password. Justin also completed a lost/stolen transfer and changed all his bank numbers and passwords. At the same time all this nonsense was going on, Justin continued to have almost 15 credit card applications (that he knows of) show up on his Credit Karma statement.

He called one of the credit bureaus and had a fraud alert placed on his account for 90 days. These kinds of alerts pop up if someone is trying to log into the account online, informing the fraudster that they would have to call in to deliver the verbal password. During this time, Justin noticed that roughly half of the requests were being processed, and he would have to call each of the corporate or credit card offices to shut the requests down. Justin became the expert on the subject, informing their offices how to handle the situation. Consistently, he knew the process better than the people handling his case. He had to first convince them to send him to their fraud department, if they had one, and then convince the fraud department to cancel the card. Cancelling the card itself was not too hard, but convincing them to cancel the hard inquiry was more difficult. Those hard inquiries gave the appearance of destroying Justin's credit file.

He said that almost all of the companies told him he had to contact the credit agency, who had previously told him it was the responsibility of the issuing company, and that they had to send a letter to take the inquiry off. That makes sense, seeing as the company was the one who sent the request in the first place. Then the company had to send Justin a letter telling him they were cancelling the request that he had asked for. He said that seven or eight times out of ten, he had to tell these companies how to do it. In the first month, he cancelled 40 or 50 credit cards.

While credit cards were the main thing, Justin remembers two instances in particular. One was a jewelry store, and the other was at a BMW dealership. He said that those people were the most freaked out and had no idea how to deal with it. Justin remembers talking to the person in charge at the dealership: "He sounded like a 40 year old frat boy, someone who is used to being in charge, but this totally freaked him out and he had no idea what to do about it."

Justin said that he found the shell address that his was changed to. Justin is an engineer. He's logical and investigative, and is lucky enough to have a flexible work schedule. He looked it up on Google Earth. He found the phone number. When he called it, unsurprisingly, all he got was an answering machine. Justin had to become the expert, and says he spent 80 hours on his identity issues for the first week. It was a full time job, and had he not had the ability to do the research, and make phone calls during working hours, who knows what would have happened. He is also assertive, and it is hard to imagine him backing down from a representative on the phone telling him what to do when he knows otherwise. For anyone who isn't like that, it's hard to imagine how they would have gotten through it.

One notable feature of Justin's story is how highly he speaks of Wells Fargo's process for handling his claim. He said they were ruthless, and their handling of the situation is why he has stuck with that bank for so long. It is good to hear that the financial institution was so helpful for Justin, and other industries could learn from their customer-focused approach.

Government, Meet Your Businesses

Inter-industry communication applies not only to corporations, but to government agencies and institutions, as well. Insurance and financial groups can learn from each other and share information, and so too can government groups. If everyone is facing the same problems, why not try to solve them together?

The government has the lowest passing rate when it comes to testing and passing software vulnerability scans, with the healthcare industry coming in close second. With 75 percent of government and 67 percent of healthcare organizations failing software vulnerability scans,[138] it seems like these two have a lot in common.

One article puts it succinctly: "While the financial sector completely overhauled its system after the most prominent attacks, like Target, healthcare has yet to make that priority shift."[139] They are right—the financial sector has made significant strides in responding to threats from identity fraud and cyber attacks, but other industries are getting left behind, industries that could learn from the steps financial companies have taken.

Not only does healthcare have the second worse software security, but they have the highest occurrence of cybersecurity mishaps of all industries.[140] These two statistics are probably not unrelated. Cyber criminals will look for weaknesses in a system and exploit them, and so knowing that healthcare has some of the worst cyber security makes that industry a real target.

Government is not exactly known for its speed and efficiency. While bureaucracy will always move at a slower pace, it is alarming to see what little is done in the time designated. The National Association of Secretaries of State (NASS) is the group of Secretaries of State that created the Business Identity Theft Task Force in 2011, if you recall, to help combat business identity theft. This emergent issue grabbed the group's attention, as they are in charge of their respective state's businesses and serve as the keepers of public business data.

Due to the Model Business Corporation Act (MBCA, the law that is not actually a law but everyone acts like a law), Secretaries of State do not have much discretion when it comes to what they publish and do not publish, and have "no authority to control who can view or gain access to state business filings, which are public record."[141] These websites serve as libraries of fraudulent opportunity, and are the building blocks for many types of business identity theft.

In 2012, NASS's task force published a white paper called *Developing State Solutions to Business Identity Theft*. The focus of their paper is "the unauthorized alteration of business records filed with the Secretary of State's office in order to carry out fraudulent acts using the identity of the affected business."[142]

The paper identifies tactics like filing false reports with Secretary of State offices and manipulating existing online records. They discuss changes of registered agent information, which then gives them the legitimacy to apply for credit accounts or loans, all tactics we have seen countless examples of. The report also points out the same issue we do when it comes to a lack of reporting: no one discusses the scope of business identity fraud, and because no one talks about it, the elephant in the room is quickly transforming into a fire-breathing dragon. The paper outlines the problem succinctly:

> To date [2012], Dun & Bradstreet has confirmed cases of business identity theft in at least 26 states. However, the exact number of business identity thefts involving altered state business records is not clear. There is currently no central repository for collecting this information... Even if the federal government did seek to do this, most states do not have a standardized method for reporting and tracking this type of crime, and businesses often fail to report it.[143]

No numbers, no place for the numbers if we had them, and businesses unwilling to report the numbers in the first place. It also mentions a lack of penalties and legislative hogties. Holding information hostage is particularly problematic in the law enforcement field. As we know, identity fraud crimes often occur over multiple state boundaries. Stepping across that invisible line makes it much more difficult for law enforcement to keep up. Boundary blindness is not only helping fraudsters, but it is keeping government agencies from initiating anti-fraud legislation.

Colorado Case Study

In 2010, Colorado's Secretary of State reached out to state businesses warning about a particular scam. Known businesses' identities were being used to create an air of validity to perpetrate business identity fraud. Using at least 35 of these businesses' identities, fraudsters opened credit cards at stores like Home Depot, Lowe's and Apple. Using the store credit cards, they purchased thousands of dollars of goods. They made at least $750,000 in fraudulent purchases at Home Depot alone.[144]

The access point for businesses' identities was, unsurprisingly, the Secretary of State's website. Much like other states, users accessing the site had free range when it came to Coloradan business information. The site allowed anyone to alter or update business records without a username or password. Also unsurprisingly, the companies' identities used to obtain credit cards were small to medium-sized businesses.

It gets confusing figuring out what is being stolen from who, so here is a hypothetical example. I log onto the Colorado Secretary of State's website, where I find Linda's Construction Company. I get all the information about the company, and I also add myself as an owner of the company. Just to be careful, I use Heather's name instead of mine. Yeah, remember her? Just like Adam's company was conned using the conned identity of a legitimate power of attorney based in New York, I could add another layer of security by using Heather's identity, making it difficult to tie it back to my actual identity.

I could steal the identity of Linda's Construction Company, and go apply for a company credit card at Home Depot in its name. Since I am applying as a business, the limit on the credit card will be much higher than if I just applied as Linda herself, or Heather for that matter. Either way, any verification by the company that issues Home Depot's credit cards (like Visa or Mastercard) will check a D-U-N-S number, which Linda presumably already has. Not seeing Heather's name on there, they might check the Secretary of State's website. If they see Heather's name listed there, it tells a compelling story.

The reason the state decided to not have usernames and passwords or any type of online security reminds us an awful lot of states' reasons for having open access to birth and death certificates: when the technology was implemented, they did not have an identity theft problem. The identity theft hole opened because the opportunity came into existence, and fraudsters filled the void. The online business registration system for Colorado came out about ten years ago and was designed to be easy to use, not to keep identity thieves out.[145]

A takeover of a business's identity can happen pretty quickly. In Adam and Corporation X's case, it took less than ten days. If the business is active, the fraudsters need to act fast before anyone in the company notices. A bigger reason why they go through at all is because the businesses are often defunct, meaning that no one is checking on them. In fact, 80 percent of the state's 356 reported identity theft victims from 2012 were delinquent or dissolved entities.[146]

Colorado has had a tough time with its business identity theft problems, but its Secretary of State is trying to do something about it. Before 2012, an email notification program was being implemented in both Colorado and Georgia. The program would automatically email anyone associated with the corporate entity whenever a change to the file was made. This is a great idea, except it has one tiny problem. You can unsubscribe from the email notifications, which means you can unsubscribe someone else from the email notifications. All you need is that business's email address. The instructions are laid out on the Secretary of State's website. It does note that an email will be sent to the previous email address noting that you have been unsubscribed, which is a great provision, except that if the business is dissolved, no one will be checking anyway.

In the spring of 2016, Larry met with the Colorado Secretary of State. Their office wanted to focus on business identity theft prevention, rather than the aftermath clean-up. A familiar sentiment for those who are used to dealing with the pay and chase model. Colorado currently has 650,000 registered companies, and they are aware that most of those businesses are very small.

The Colorado Secretary of State's office employed two tactics when they realized they had a problem with business identity fraud. The first was email notifications, which sends notices to a business's owner when a new form has been filed, or if the business's status changes.

The second tactic is called Secure Business Filing, which ties that email address and password to a business record. Once the account is created, forms affecting a business cannot be filed without the appropriate email address and password.

Their email notification program has been very successful, with 94.3 percent of businesses receiving emails. The Secure Business Filing (with multi-factor authentication) has been less successful. As of September 27, 2017, only 66,171 out of 665,780 good standing businesses had secure business filing accounts. That means only 9.93 percent of businesses will get notifications. Despite the number of Colorado businesses showing continual growth, the amount of Secure Business Filings has plateaued around 10 percent.

So, if a business is registered as having a Secure Business Filing—which is separate from having a registered email—the account requires an email address and password to gain access and change any forms. While the addition of the email alert system and the passwords to all registered business accounts is a good idea in theory, business owners have to register to be a part of the program. They are not included automatically.

One prediction was that 97 percent of the state's active businesses would be signed up for the secure business filings program by January 2012.[147] Despite heavy outreach, only 20 percent of the businesses have gone back and re-registered their secure business filings[148] within the system which would allow them to receive the alerts. Those numbers do not bode well for a program that cost almost $361,000 to implement, though it does not seem like there is much more for the Secretary of State's office to do. You can bring a horse to water, but you can't make it register for state programs.

After hosting a forum to figure out public perception, participants noted the potential usefulness of verification programs like this one, in confirming the identity of the individuals attempting to make changes to business records.[149] Unfortunately, these systems only work if business owners do their part by participating in them.

I figured that the best way to test this was through a hands-on approach. I asked two friends who own a business in Colorado (and who have a pioneering, anything-for-science spirit) if I could try to steal it. These people are young and their business is still active, so it's not exactly the target market. Younger people are more familiar with email and more willing to sign up for notifications and register an email address. Since the business is active, they would receive email notifications if I tried to mess with it. The password protection worked. While I was able to subscribe to receive emails with my email address, I could not access their account since it was password protected, though it was easy to find lots of information about their business that they may not have wanted out in the public sphere.

What was scarier was finding a delinquent business that I will keep anonymous, which was not password protected. Since Colorado's measure requires the businesses to go back and re-register for notifications, most businesses do not have a password-protected account. When I clicked on the 'change address' and 'change registered agent' tabs, the forms popped right up. Since I did not know this individual, I resisted the urge to see if I could do it. The difference is that when I clicked those same buttons for my friends' password protected account, I was not able to even view the forms. This leads one to believe it would be pretty easy to change the address or registered agent for this anonymous business, and we all know what can happen after that.

In a second meeting with the Secretary of State's staff, Larry discussed the damaging factors of the way the business filing system works. First, all submitted filings for business creation and modifications are A) self-reported and B) public record. No information aside from name and address are kept on business owners or registered agents. If someone is fabricating information, they do not even have to have sensitive data to do it. Since the Secretary of State's office does not have the authority to verify information that is submitted in agreement with their "ministerial" role as prescribed by the Model Business Corporation Act, it is almost a moot point. They are not allowed to verify data, and even if they could, that data is fallible. Larry also learned that companies outside of the United States that need to conduct business in Colorado have to go through the Secretary of State's office. In other words, there are companies from across the globe that register, with no verification.

We have discussed boundary blindness and how it enables fraud to evaporate across state boundaries. One of the points the Secretary of State's staff brought up to Larry is how the identity theft task force they created can only investigate within Colorado state lines, otherwise they have to get Colorado state law enforcement involved, who then has to reach out to the other state's law enforcement. Often, the amount it would cost to investigate across state borders is less than the amount stolen.

Fraudsters seem to know that if they combine two factors, crossing state borders and keeping the stolen amounts below $50,000-$100,000, it will minimize their chances of being pursued. Often, criminals will limit the amount they steal to $50,000, but will do that with multiple companies. A federal approach will not combine two $50,000 crimes and total it at $100,000, but rather they look at each amount individually. Ralph Gagliardi, the Agent in Charge for the ID Theft & Fraud unit for the Colorado Bureau of Investigation, says he sees a repeating pattern: "Often they will do the same scam on twenty or thirty companies each for $50,000, and walk away with millions after several other iterations."

"Large retail companies who issue credit are popular targets," Mr. Gagliardi said. "These hijacked businesses will use their good credit to purchase thousands of cell phones. These phones are then shipped to an address that the fraudster controls or changed curing the original hijacking. The fraudsters immediately reship the phones outside of the United States. Depending on the newness of the phone they can get triple the amount for the phone netting approximately $1000 to $3000 per phone."

The Colorado Bureau of Investigation is looking into an emerging scam involving online retailer's fraud. Individuals can register a business they created or modified on the Colorado's Secretary of State website, which allows foreign entities access to Colorado business information. These businesses can then be used for fraudulent customer purchase ratings on the retailers site increasing their exposure thus allowing the fraudsters to sell counterfeit merchandise and conduct other schemes. While the office is still unsure exactly what is being shipped, they are confident that the way these businesses appear to be operating is fraudulent. In particular, they are noticing the sheer number of businesses created in unlikely locations, like empty parking lots.[150]

Time Management: the Beauty of Efficiency

Not to hold up the financial sector as the golden child of fighting fraud, but they are held so accountable by everyone in their industry, it makes it hard for other groups to compete. We have all accepted that government, especially the federal government, operates at a glacial pace. People love to harp on government for how long it takes to get anything done, but a big way around that is minimizing the things they have to do. With front-end solutions, like abandoning pay and chase for a model that will avoid improper payments in the first place, we can immediately cut out a whole slew of things for government agencies to do.

Let's get back to NASS, the group of Secretaries of State that banded together to fight evil and foster cooperation in the development of public policy. Now that you have gotten a close look at one Secretary of State's office, let's look at the task force that NASS developed specifically to deal with business identity fraud on a national level.

After learning about business identity fraud in 2012—how it was affecting businesses and the different ways it was perpetrated—NASS created a list of recommendations for the Secretaries of State to take home with them, like the worst gift bags ever. We are going to take a look at their 2012 recommendations, discuss what we think of them, and see how they have been implemented, if at all, in the last six years.

Recommendation #1: Establish a Statewide Task Force on Business Identity Theft

When in doubt, make a task force. This is not a terrible idea, though it is hard to defend government administration when a task force's first move is to replicate itself like it is performing bureaucratic meiosis. NASS throws out some ideas for what the task forces could do, like reviewing local law to see where business identity theft could get through and conducting outreach to help educate businesses about the problem.

My personal favorite idea from this recommendation is to establish a Business Identity Theft Awareness Day, Week, or Month, or to incorporate it into a pre-existing celebratory month, such as the federal government's National Cyber Security Awareness Month in October. I don't think the committee took into consideration that October is already Breast Cancer Awareness Month, Polish American Heritage Month, National Orthodontic Health Month, and Vegetarian Awareness Month, but that's another story.

Awareness is key, but devoting a sizable portion of the

marketing budget to a month few people will pay attention to, and a whole lot of marketing budget will disappear into seems a little silly. Looking at the pens on my desk, I see that they're all contrived to help spread awareness or advertise for something, including "The Painted Buffalo Inn" (never been there, do not intend on going) and the "Great Plains National Security Education Consortium (no idea what that is). My point being that pens, mugs, and months are small, ineffective, and make outsiders question what organizations spend their money on. Government already lives under a microscope of public outrage at their spending choices. Why make it worse?

If your goal is to spread legitimate information, put together a fact sheet and focus on distribution. In fact, NASS did put together a fact sheet. It was succinct and it's web page was a lovely light salmon color. When I tried to find the analytics on how many times the page has been viewed or downloaded, I was told that NASS was completely redoing their site, so the website analytics were nonexistent. Make of that what you will.

Offer incentives to businesses for showing up to meetings that discuss business identity fraud. Businesses will pay attention if they realize they could lose money. Actual information beats National Cyber Security/Breast Cancer/Polish American Heritage/National Orthodontic Health/Vegetarian Awareness Month any day.

Recommendation #2: Develop a State Action Plan

This recommendation is pretty good. It reads, "Secretaries of State need a clear plan of action in dealing with business identity theft, including a thorough understanding of the laws and policies that impact the state's abilities to fight this type of crime." Kind of thought we covered that in the first recommendation, but sure, it is pretty major so it deserves its own bullet point. They mean legislative action, staff training, data collection and management, and data sharing. I dare say, the phrase "consider sharing information with other states" is even mentioned.

They discuss developing a national repository that includes information about business identity theft victims. It also talks about the FTC's clearinghouse that contains all identity theft-related complaints. NASS identifies that keeping track of data like this is helpful to "identify other agencies involved in investigations, and spot identity theft trends and patterns."[151] That is as close a call for front-end, preventative measures as we can hope for.

Recommendation #3: Establish a Notification Process for Businesses in Your State

Informing business owners of changes to their records is a good idea. If I, or someone else, logs onto my email account from a different computer than I usually use, the platform automatically sends me a notification. What Colorado, (and now other states) has implemented operates the same way. The problem, is getting the state's businesses to adhere to the recommendations. Offering incentives to businesses to sign up for these protections would most likely increase the rate of participation.

Recommendation #4: Establish Clear Steps for Victim Assistance and Education

Do not just leave victims to flounder and figure it out on their own. NASS recommends contacting law enforcement, banks, and credit providers, placing fraud alerts on accounts, and requesting copies of documentation used to fraudulently open accounts. All this information, and where to start, could be included on a subset of the Secretary of State's website. Fortunately, many states have done this.

Recommendation #5: Conduct Outreach to Raise Awareness and Urge Preventive Action

We already covered outreach, and how labeled pens are not a viable option, but urging preventative action is a good idea. Preventative action includes regularly monitoring your filings, even if the business is closed, and monitoring your credit reports and billing information—all things individuals should already be doing for their own records. Taking the same precautions with businesses as one would with their own personal information and finances is one of the best ways to protect against business identity fraud.

NASS's recommendations are from 2012, and both individual and business identity fraud have increased since then. Javelin's 2017 Identity Fraud Study surveyed 5,028 consumers, and claimed that 2016 saw 15.4 million victims of identity fraud, and almost $16 billion in losses.[152] Did those 2012 recommendations help at all? With fraud thriving, how effective has NASS been at spreading awareness about business identity fraud?

In February of 2017, NASS put out a paper called "NASS Business Identity Theft Task Force: Findings and Suggestions for States." Five years after the Business Identity Theft Task

Force's first meeting, the task force took some surveys and made suggestions based on their findings. They found that 83 percent of the Secretary of State offices who responded to the survey are not tracking business identity theft complaints. Sixty-one percent said "they could not provide data on the number of business identity theft complaints that were received in the past year."

Remember that "keeping track of business identity theft complaints and victims" fell under recommendation number two from the list NASS created five years ago. NASS keeps making suggestions, but the suggestions only work if the Secretaries of State implement them and businesses follow them. They also recommend providing "clear remedies for remediation of fraudulent filings," which leads me to believe that 'Recommendation #4: Establish Clear Steps for Victim Assistance and Education' also never happened.

NASS cites that a third of Secretary of State respondents noted that they have no specific laws to address business identity theft. Eighty-three percent of states still do not track business identity theft data. Twenty-nine percent of states have no relevant business identity theft legislation.

NASS as an organization is a great idea. Making a point to invest government funds into dealing with business identity theft is important, but NASS as an organization cannot make anyone do anything. The individual Secretaries of State are the ones that have to do something. They cannot just come home from the latest NASS meeting and do literally nothing with what they have learned. Showing up is not enough, they have to actually implement the changes they discuss.

This is not to say that all Secretaries of State are just going home and twiddling their thumbs. States like North Carolina and Colorado have assigned identifiers to use across all state agencies. That means when a business comes up in any state database, they could see a blue flag, meaning they need to check with a supervisor before proceeding, or a red flag, meaning they should cease any additional filings. Front-end solutions like that start to take the Secretary of State's office out of the solely ministerial role they have been deigned to.

Everyone knows government does not move quickly, but there is no reason it cannot. If government can take a page out of the financial sector's book, and make adapting to fraudulent tactics a priority, it could make a significant difference in how frauds do or do not develop. Utilizing technology and embracing the fact that government still has much to learn could really help them stay accountable to their constituents.

NASS was not even the first government task force to be dedicated to fighting fraud. In 2006, the President's Identity Theft Task Force was established, and in 2007, they made such legislative recommendations as "amending the identity theft and aggravated identity theft statutes so that thieves who misappropriate the identities of corporations and organizations—and not just the identities of individuals—can be prosecuted," and "enhanced sentences imposed on identity thieves whose actions affect multiple victims."[153] Heavy stuff.

This pre-NASS Business Identity Theft Task Force foresaw the issue of business identity fraud in 2006. Some states, like California, took the hint and changed the language in their legislation so businesses can be included as potential victims. Florida made the change in 2015,[154] but even now, few states have enacted legislation to protect businesses' identities.[155]

At the Speed of Molasses

The Criminal Use of False Identification is a book published by the Department of Justice 1976 that outlines problems with matching birth and death records, fraudulent uses of real birth certificates, and identity theft. Sound familiar? This book provides just a taste of how slowly legislative changes are made, if they ever are, and how procrastinating on these issues can harm agencies and industries. It is absurd that in 1976, every single one of the problems we discussed in the first section of this book were already addressed by the Department of Justice.

The book identifies issues with matching, pricing, and identity theft: "It is well known that false birth certificates are obtained (and subsequently false identities established) by applying for the birth certificate of a deceased person."[156] *It is well known.* If it was well known in 1976, why has very little been done about it until 2010? Why then, after knowing how serious of a problem business identity theft is, and identifying how to work on it, has a good portion of Secretaries of State done little to nothing to combat it?

Two decades later, a paper called "Reinventing Vital Statistics" called out the sloth-like pace at which data moved in government in comparison to retailers:

> Financial markets are so thoroughly wired that they generate data almost instantaneously. Births, deaths, marriages, and divorces are hardly less important to society than the sale of toothpaste, dispatch of an overnight letter, or purchase of a share of stock. Yet data on vital events take a lot longer to travel from their point of origin to relevant decision makers. The

> Federal Government reports national birth and mortality statistics with a lag of 12 to 15 months after the close of a year.

From 1976 to 1995 to 2017, it does not appear that the pace has picked up. Maybe it feels odd to bring up birth and death certificates again, but by its nature, the themes of fraud and abuse are pervasive in every industry. Candidly discussing business identity fraud would allow for a more open discussion, more studies to be done, and more tools to be put in place to mitigate the problem. If we all pretend there is no problem, obviously it will take longer for it to get fixed across industries.

The same goes for government. If the Department of Justice is highlighting problems in the 1970s, and all the recommendations and policies they suggest are ignored, how can we expect the problems go away? A more streamlined process and strong initiatives to fix problems in a timely fashion would go a long way.

Data Disposal

Business and government alike have a serious problem with something none of us can see, but that is very sensitive: data. Here's the bottom line: the more sensitive data you have, the more sensitive data you have to protect, and usually you have to pay to protect it. In many cases, industries collect far more data than they actually need. (How many times have we mentioned that you do not have to put down your Social Security number at the doctor's office?)

One thing both government and businesses can work on, from healthcare to finance to insurance, is limiting how much data they collect. This is something larger businesses are really starting to understand. With data breach headlines becoming commonplace, it is easy to forget how detrimental losing customer trust can be. When companies like Target and TJ Maxx experience data breaches, of course they do not want to talk about it, because their customers lose faith in that company when it is discussed.

Sean McCleskey from the Identity Institute in Texas advises a couple of different strategies to help deal with issues surrounding business identity fraud: "Organizations need to build data security practices into their culture." While he recognizes that there is no perfect checklist out there to help businesses create an airtight anti-fraud strategy, he says that would not be a bad place to start. "An eternal truth of fraud is that there will always be a fraudster there to take advantage of a system. If you create a checklist of items, that is a good place to start. It will give you a general direction that will help you discover what is not on the checklist, and thus what should be."

Some things that would be first on the list? Not asking for data you don't need in the first place. Second? Making transparency and fraudulent risks a priority. Businesses cannot be expected to implement changes if they do not take steps to make educating employees about potential risks a primary concern.

If you have too much data, think of it as spring cleaning for the cloud. Most states actually have data disposal laws for companies, meaning that after a certain length of time, you have to dispose of your customers' Personal Identifying Information.

In 2015, the Secure ID Coalition performed a study where they took an in-depth look at 10 different states' identity policies. The 10 they picked were because of their reputation as technology leaders and because they prioritized states with larger populations. In one section, they awarded three points to each state if they had laws that require businesses to destroy PII once it has been used. Zero points were awarded to states without data disposal laws. Think of 'three' as 'yes' and 'zero' as 'no' and examine the following chart:

State	Data Disposal Laws for Businesses	Data Disposal Laws for Government
California	3	0
Illinois	3	3
Michigan	3	3
New York	3	0
Utah	3	0
Florida	3	0
North Carolina	3	0
Texas	3	0
Virginia	0	0
Pennsylvania	0	0

This chart is a reproduction of one featured in the "State Secure Identity Practices and Policies in 2015" paper. The states that feature a 'three' mandate data disposal, while a 'zero' shows they do not mandate data disposal.

Only two out of the 10 states did not have data disposal laws for their businesses. If their businesses record PII, by law they need to dispose of it. When it comes to government, however, the reverse statistic is true. Only two out of 10 states require their governments to dispose of PII after it has been used. Interesting to examine, especially when these states are considered industry leaders who have "instituted innovative identity initiatives," despite their governments falling behind.

Looking at these charts may give you the impression that governments are immune to having their constituents' PII compromised. Except for that time in 2015, when hackers exposed 21.5 million people over government computers, including Social Security numbers and fingerprints.[157] Or that time in 2016, when the Russians felt passionately about the United States election results.

Why are our elected officials held to a lesser standard than our businesses, especially when it comes to American citizens' personal information? If our government agencies and institutions were held as accountable to their citizens as businesses are to their customers, it would certainly help reduce fraudulent costs, and maybe someday, I would hear a story from a friend about how the IRS was just as helpful and easy to work with as Wells Fargo.

7

Where Identities are Headed

Put succinctly, attackers and defenders are in an IS security arms race, where the brunt of the costs are borne by the defenders, and the attackers are winning by almost every measure.
— Sean Zadig, Understanding the Impact of Hacker Innovation upon IS Security Countermeasures, 2017

I was recently on a flight from Salt Lake City to Seattle, and having not flown for a while, was surprised to find a USB port built into the back of every seat. Looking across the aisle, I saw that almost every passenger had already plugged their phones in. It occurred to me how funny it was to see every person with their own cable and their own phone plugged into their own USB port. While I am no technology expert, it seemed like a lot of individual parts in order for every person to accomplish the same goal: charging their phone.

Since Wi-Fi became ubiquitous, each individual person does not need to bring their own ethernet cable with them wherever they go—they just need a password and they can access the same wireless network everyone else is on. Wi-Fi can be seen as a front-end solution to everyone wanting to access the internet, whereas individual ethernet cords, or power ports in this case, are more akin to band-aid solutions. Overarching solutions address the original problem; band-aid solutions address every aspect of the problem but the source, every side effect that crops up.

Government agencies and businesses have adopted band-aid solutions to deal with fraud in their organizations in the past, often by accident. Fraud involving identities is a fairly new trend, and so many of the older technologies we still rely on never accounted for crimes of the future. Band-aid solutions will always play catch up rather than stopping the problem before it starts.

On July 1, 2010, the Puerto Rican government enacted Law 191, which invalidated all birth certificates issued before that date. The law required all individuals born in Puerto Rico to get a new version of their birth certificate. These new certificates incorporated "technology to limit the possibility of document forgery."[158]

A big reason why Puerto Rico in particular had such a problem with birth certificates is that it was common practice for agencies to retain a copy of a real birth certificate whenever proof of identity was required. This meant that schools, hospitals, insurance companies, and sports coaches typically had tons of real, valid birth certificates on file.

As we noted in the last section with regard to large amounts of data, the more information you have, the more information you have to securely store. In most cases, the Puerto Rican agencies and individuals who maintained these birth certificates did not do so with adequate protection. There were many cases of birth certificates being stolen and consequently sold in order to obtain passports, licenses, and government benefits.

A statistic cited by the Puerto Rico Birth Certificates Law

Fact Sheet states that "approximately 40 percent of the passport fraud cases investigated by the DOS Diplomatic Security Services in recent years involved birth certificates of people born in Puerto Rico." Forty percent is a staggering amount, and compels some legislative change.

Let's first look at the good parts of this new legislation. The law establishes that new birth certificates will feature technology like counterfeit proof paper with special seals and motifs to stop counterfeiters from duplicating the documents to limit the possibility of document forgery.[159]

The other good component of the new law states that "no public or private entity... may retain an original copy of a Puerto Rico issued birth certificate." It goes on to state that these agencies, including government agencies and private employers, can inspect and copy birth certificates, "but cannot retain the original under any circumstances." This is an excellent provision to this law, and aligns nicely with the idea of reducing how much data an organization stores unnecessarily.

While in the States it would seem highly unusual for any agency or individual to ask to keep your birth certificate, in Puerto Rico, it has been public practice. Individuals or agencies that violate this new law are subject to a criminal misdemeanor and could potentially be held liable for the damages incurred due to the violation. Enacting laws like this bolsters the idea of limiting data, or in this case, birth certificates.

There are also not-so-great parts of this law. By invalidating all birth certificates generated before 2010, a logistical nightmare is born. Spreading the word about the new law so Puerto Ricans living within the country and abroad are aware of the changes, making sure people can order new birth certificates, and preventing any new kinds of fraud that might pop up, is a tall order. In Massachusetts, both state and federal agencies have preemptively enacted policies to ensure that Puerto Rican-born citizens are not denied access to benefits like food stamps, Medicaid,[160] and other services in the wake of having to obtain a new form of identification. It also complicates matters when applying for a driver's license.

Considering all the complications, did this actually work? Did invalidating old birth certificates reduce identity fraud in Puerto Rico? Unfortunately, that does not seem to be the case. Especially since you can still go online to the ordering platform and obtain a birth certificate if you are a parent, legal guardian, heir, or authorized by the court. The states or territories that adopt this policy still only have the security of a drop down menu.

An investigation performed by the Miami Herald noted that "half a dozen cases have popped up in Miami federal court involving defendants who have illegally procured Puerto Rican birth certificates to obtain passports or driver licenses."[161] In several of those cases, large sums were paid, typically around $1,500 to $2,500 in order to obtain Puerto Rican birth certificates, driver's licenses, or passports. The article notes that "While officials in 2013 [when the law was put into effect] did not say Puerto Rican birth certificate scams would end under the new system, the general expectation was that such ID theft would be more difficult."[162] It is the general expectation that when laws are put into place, they will fix the problem at hand.

A Watchdog article points out that while there are 4.9 million Puerto Ricans living in the United States, and another 3.7 million in Puerto Rico, there are not enough Puerto Ricans that exist to support the numbers.[163] Roughly 100 Puerto Ricans arrive in the United States every day. The only explanation for it is that non-Puerto Ricans using Puerto Rican birth certificates are actually the ones arriving in the United States.

Elizabeth Cuevas-Neunder, a Republican candidate for governor of Puerto Rico back in 2014 noted, "The law approved in 2010 doesn't work, because if you already have a (fraudulent) birth certificate you can renew it online or by email. Nobody is certifying that you are the person you say you are."

Parts of this law, such as not allowing agencies to retain documents, are excellent front-end solutions. Other aspects, such as invalidating everyone's birth certificates while still allowing copies to be ordered online without any additional checks as to who is ordering, do absolutely nothing but inconvenience everyone who has to go through this process.

By invalidating government-issue documents, it also sends a strong message that birth certificates from that government ought to be suspect. Puerto Ricans have already had a difficult time both proving that they are native-born citizens, and that their birth certificate is just as valid as any other. In New Hampshire, Puerto Rican passports had to go to a special office in order to apply for a driver's license. People applying for driver's licenses in other states have already had birth certificates rejected as invalid. There are ways of keeping documents secure that does not involve marginalizing groups of people.

Securing documents is not a new idea. Between the pages of the 1976 Department of Justice book, the Office of the Inspector General's report in 2000, and the plethora of works that came after 9/11, the same ideas have been repeated over and over again: linking birth and death certificates, closing open states' document policies, standardizing documents so we do not have thousands of variations in existence, and getting full access of databases that contain information. These ideas have been so often repeated that when I read a government issue document on the topic, all I do is nod in agreement. This information is no longer shocking; how long it has taken to implement changes to secure our personal information is shocking.

Take, for example, the REAL ID Act, which has still not been nationally implemented. We talked about the REAL ID Act in our first book, and have had time to publish a second book before it has been implemented. The Department of Homeland Security's website crows, "It has been 12 years since the REAL ID Act was passed and half of all the states have already met the REAL ID minimum standards." To have only half of the states meeting the minimum standards after 12 years is not something to be proud of. Accepting the "business as usual" maxim for bureaucratic pace is not going to work for this issue. Fraudsters move fast, they evolve, they adapt.

The information I provided in Section I, including all the documentation I needed to create Heather and her fraudulent identity, would still be enough to circumvent all the precautions taken with the REAL ID Act. I could have a REAL ID-compliant license with my picture and Heather's information on it. The slower initiatives move, the more pointless they become. The 'minimum security requirements' mandated by the REAL ID Act of 2005, which Heather's identity can surpass, could be circumvented if birth certificates followed a recommendation from a 2015 paper by The Document Security Alliance: raising the physical security of the birth certificate to equal the security of the documents it breeds.[164]

That could be achieved by any kind of marker, such as an evolving photograph or biometric data tying the document to the person it represents. By having nothing to link the birth certificate to the owner, it leaves an enormous window open. Creating actual change, both legislative and within agencies' own cultural climates, is the only way to decrease the amount of identity fraud that occurs. Front-end solutions are the only way forward.

What Can Technology Do for Us?

Odds are, you have seen an attempt at a front-end solution in your own life. Does your credit card now have a chip on it? The thing the exhausted cashier tells you to stick into the machine rather than swiping it? You leave it there for four to eighty minutes, and then suddenly the machine starts harassing you to get your card out with great urgency?

These chips, technically an EMV card (standing for Europay, MasterCard, Visa, the three technologies that created the benchmark technology) suddenly appeared in a wave after 2015. They were not, as you might expect, the consequence of legislative change, but because of a liability shift. Meaning, if there was a hack, whichever entity (the bank or the store) had the lesser technology would be responsible for reparations. So stores all the way out to Wyoming started implementing chip readers.

Chip technology works through encryption. With a regular magnetic stripe credit card, the card reader simply reads the information off the card, like the card number, expiration date, etc. This is how skimmer frauds took off.[†] With encryption, every interaction a card has with a chip reader fundamentally alters the card. Think of it as a conversation that the chip then has a record of built right in. The dynamic data created by these so-called conversations means that it cannot be recreated. There is a unique code that is generated for each transaction—which means if someone steals that and tries to duplicate it, the card will simply be denied.

The United States is actually one of the last major countries to start using EMV technology. It is already prevalent in Europe and most of Asia. Counterfeit card rates have been falling in other countries since the implementation of EMV cards, which corresponded closely with the rise of it in the United States.[165] Hopefully once the program is fully up and running, the United States will see similar declines in fraudulent credit card transactions. The financial sector's timetable for this implementation gives the government agencies a lot less of a leg to stand on when it comes to time management

While it does seem that EMV cards have put a damper on some kinds of credit card fraud, it does not fix the entire problem. Remember, fraud is creative. There is no telling how

---

[†] Fraudsters created devices that could slip over credit card readers at gas pumps and ATMs, and simply read off the data on the magnetic stripe and store it. That way, fraudsters have all the credit card information necessary to make purchases. Since nothing is sacred and fraud is an ever-adapting beast, there is also 'shimmer' fraud, where the device acts as a literal shim, blocking the chip on the card from meeting with the chip reader and records the data on the chip. Shimmers can't fabricate a chip-based card, but that data can be transferred to act like a magnetic stripe card, but only if the bank card issuer neglects to check the CVV when authorizing a transaction. The 'dynamic CVV' is the other component that makes chip cards special, though those too can be compromised. As long as that "conversation" is happening between the chip and the institution, the card information cannot be copied. For more information see Brian Krebs' article "ATM 'Shimmers' Target Chip-Based Cards" on his website, Krebs on Security.

fraud will evolve and change in the next few years and the next few decades. It is an endlessly creative market, and while EMV cards seem pretty solid at the moment, there will surely come a day when they are no longer a useful solution.

Already, there are problems with them. EMV cards are not any more secure when it comes to online purchases. Which fraudsters will surely shift to if they have not already. Online shopping means there is no interaction with the chip, which means none of those secure conversations between financial institution and card ever happen. These kinds of "card not present" frauds have skyrocketed 186 percent from 2004 to 2012 in the United Kingdom. If we can only hope to follow Europe's example, we should have the foresight to see and plan for the next shoe to drop.[166]

There will always be problems with fraud. There will always be people who try to game the system. Always. There will also always be people who are just as innovative on the other side. Utilizing the genius that surely exists within individuals who do not want to commit fraud is critical. Coming up with new ideas, like revamping how credit cards function, will be central.

The 1995 "Reinventing Vital Statistics" paper, which said vital statistics data should move faster than the sale of toothpaste, discussed the use of an electronic birth and death certificate.[167] From that point, we can draw the conclusion that that hospitals were still in the process of switching over from a paper to an electric system in the mid 1990s. Now that our systems are electronic, it would be infinitely easier to make changes and start adding more security features in a widespread fashion.

With the latest iPhones using biometrics fingerprints as locks, why couldn't we use biometrics within a birth certificate?

In fact, there is already an existing United States patent for a biometric birth certificate. In 2011, a gentleman from Iowa filed a patent for a birth certificate that holds onto a portion of an individual's DNA:

> A DNA birth certificate has a DNA sample portion to securely hold a biological sample for use in positively identifying an individual. It can be used to keep track of babies to prevent hospital mix-ups or in child stealing cases. Additionally, the certificate can be an ID or even a driver's license. The DNA sample portion holds one or more containers that allow the user to place a biological sample such as a hair or nail clipping within to provide the DNA if needed.[168]

A more recent call for a biometric birth certificates came in 2017, when a Hawaii paper explained that while biometrics have been utilized in other forms of identification, birth certificates have been left in the dust. Their argument is that the DNA sequence used to identify infants at birth is the one thing that remains static through our lifetimes, and it can certainly be linked to our documents. The paper claims that incorporating biometrics into birth certificates "allows for a person's life to have a definitive beginning for purposes of documentation and record keeping."[169]

It is shocking that birth certificates are so far behind in the discussion of document security. While passports and other documents have an impressive array of security functions—some including biometrics, and credit cards can include a picture of your face—birth certificates contain virtually none. They are flimsy excuses for identity verifiers, and while people in government seem hesitant to safeguard them, fraudsters have no problem taking advantage of their weaknesses.

Biometrics can be used for two things: authentication and identification. Authentication is what the iPhone fingerprint entails: it matches an image within the device. For identification, an image has to be run against a database filled with lots of other images to sift out which one it matches out of potentially hundreds of thousands.[170]

Individuals nervous about biometrics have already lost the argument. Our body parts are inherently public, in that someone can see our face, and take a photo of it from miles away. Police have been using fingerprint identification for decades. The following agencies have already started, and some have been embroiled in legal battles because of it. An article from Wired.com sums it up:

> The FBI has been building a biometric recognition database that it hoped to have filled with 52 million facial images by 2015, with thousands more images added every month. The Department of Homeland Security is working with U.S. Customs and Border Patrol to add iris scans and 170 million foreigner fingerprints to the FBI's national database. And local police departments are also in on the biometrics game.[171]

Fears of overarching government and Orwellian controls creep into arguments against better security, but in truth, those kinds of controls are already here. Facebook has essentially the largest facial recognition database in existence, with over 350 million tagged photos uploaded every day. Users do the work of categorizing their images for Facebook. The social media mogul has an agreement with the FTC that they must receive "affirmative express consent" before bypassing a user's privacy controls. That "affirmative express consent" refers to the fine print of Terms and Conditions that normally people do not read when they are signing their privacy away.

Currently, tech companies, social media sites, and online shopping is like the Wild West. These companies have been self-regulating, and government has been a little behind in keeping up. Without even realizing it, we already have biometrics on file, not necessarily with government agencies, but with businesses, though government is not far off. In 48 states, it is legal to take pictures of you without your knowledge for identification purposes.

These systems are not immune, either. Biometrics is not a guarantee of a fraud-free future: fingerprint scanners have been hacked using play-doh and gummy bears, and it gets scary imagining how far people will go to obtain body-specific identifiers. Since everyone else seems to have biometric data, it seems silly to bar the government from using it for positive reasons, but regulating those reasons would be critical.[172]

The reasons to not have biometrics are the same as the arguments both for and against a national ID card. Many Americans are vehemently against any form of a national ID card, since it interferes with their ideas of an unfettered and unregulated freedom. As we already pointed out, we basically do have a national ID card, in the form our Social Security number. This number serves the same purpose as a national ID card, but it leaves vulnerabilities all over the place. Having a national ID card with legitimate security measures in place would definitely make it harder and more complicated to commit identity fraud.

Other technologies like Artificial Intelligence (AI) are starting to be implemented as well. AI, also called 'machine learning,' tracks users' patterns, such as how often you go to the grocery store or fill up your gas tank. These human patterns help these technologies understand what is a 'usual' transaction and what isn't, which makes catching fraud a little more manageable. AI has not been easy to implement, since it is so complicated to create an algorithm for people's habits and to check those algorithms in real time.[173] Creating foolproof AI is no easy task either, but this emergent technology could be a significant asset in the fight against fraud. Stopping fraud before it happens is the definition of a front-end solution. It means no catch up, no pay and chase, and no law enforcement involvement, saving everyone time, money, and heartache.

Employing AI could mean a world of new possibilities. It could check benefits applications in real time for local, state, and federal agencies, it could assist in strengthening verifying documents, and it could have promising implications for protecting against business identity fraud, as well. If fraud is determined by how creative the perpetrators are, the individuals implementing solutions will have to be twice as creative.

Business, Meet Your Government

One study found that business identity fraud costs the global business community $221 billion every year.[174] That is too much money to be losing, especially knowing that a big percentage of the businesses affected are not the large well-known types that can swallow these costs and move on. Most of these businesses are small to mid-sized, and cannot afford to be losing so much, if anything, to fraud. Behind each of these small businesses is a business owner who has a lot to lose.

The recommendations for how to protect a business from identity fraud, like the tactics used to defraud a business, are similar. Limit how much information you have to store by only taking in what your company needs. Additionally, if you do need sensitive information, destroy it after you are finished using it. You can't have information stolen from you that you no longer have.

When it comes to paper documents, keep them in a locked filing cabinet. I have seen plenty of filing cabinets, unlocked, with surprising amounts of valuable information inside. Electronic documents should be equally protected. This may mean having passwords and encryption, the electronic equivalent to locking a filing cabinet. Your passwords should not all be the same, should not be your birthday, nor the name of your pet. Data encryption uses an algorithm to convert your normal text into a mush of indecipherable characters, meaning no one aside from the intended parties and talented hackers can read it.

For a small to mid-sized businesses, digital fraud prevention solutions ought to be common-sense. Sign up for email notifications for your different accounts. Then, if someone attempts to change the password to your company bank account, you will be emailed as soon as it happens. Monitor your accounts in case of unwarranted changes or charges. If your state offers it, sign up for email alerts from your Secretary of State's office. If someone adds themselves as a member, or a registered agent in your stead, then you will at least be aware of it. In many of the cases we studied, swift action was the only thing standing between these businesses and major financial loss.

For individual businesses, the steps to take are small and simple. For government, they are more nebulous. Data collection and management come in here, too. As we saw in the data tables in the previous chapter, a good portion of states require businesses to dispose of sensitive data, while almost none require their governments to do the same.

Sharing data that is helpful, between states (and yes, then disposing of it) is also a good move. Training law enforcement to recognize business identity fraud is critical. When individual identity theft, and now business identity theft, was an emerging crime, law enforcement had no idea how to respond to those calls. "Someone stole your identity? Sure. Just pay your bills sir." Click. Utilizing security experts to explain the emergent frauds, and further collaborating to figure out the best way to implement those changes is one of the best things local governments can do.

Unfortunately, these types of changes are far more difficult than it is for one business to put a lock on their filing cabinet and update their passwords more regularly. The likelihood of these systematic changes occurring will depend on the government's understanding of how big a problem they are facing, how serious it is, and how much they stand to lose. The question remains: what will it take for them to see how big of a threat this is?

Nothing is more indicative of this than when Larry attended the NASS conference, supposedly held to discuss business identity fraud, the problem they created a task force for years ago and have not done a whole lot for since. Instead, NASS devoted their entire conference to talking about voter fraud, a problem that has been debunked so many times by actual research it is starting to feel like a fraud itself, disguising its uglier face known as voter suppression.

Hysterical media claims that the dead people voting in large numbers are interesting, because while dead people are not actually doing that, they are procuring benefits and drivers' licenses. The Pew Charitable Trust report, which is often cited for that unsubstantiated NASS claim, says there are 1.8 million deceased individuals who are listed as voters in this country. It does not say that those listed are actually voting. Pew was using that number to show the need to upgrade the voter registration system. This need is, in part, a reaction to the Help America Vote Act of 2002, which was put in place to encourage states to improve the accuracy of registration lists and audit their election results. This act had some states comparing their voter lists to the Social Security Death Index, whose information is supplied by the Partial Death Master File, which we now know is highly inaccurate.

This comparison to the SSA's death data supposedly turned up a link between voters and the deceased, but according to Factcheck.org, "almost all of those turned out to be due to clerical errors or as a result of people who legally voted via absentee ballots or the early voting process but later died before Election Day."

According to a report produced by the Brennan Center for Justice, "Much of the misinformation about "dead people voting" is due to 'flawed matches from one place (death records) to another (voter roles)." In general, reports of dead people voting are either unfounded, greatly exaggerated, or based on clerical errors that did not actually affect voting. FactCheck.

org has been countlessly sourced for this type of data, like the following:

> In 2012, a team of students led by the Walter Cronkite School of Journalism and Mass Communication at Arizona State University analyzed 2,068 alleged election-fraud cases since 2000 and concluded that "while fraud has occurred, the rate is infinitesimal, and in-person voter impersonation on Election Day, which prompted 37 state legislatures to enact or consider tough voter ID laws, is virtually non-existent."[175]

It is infuriating to think that 37 states jumped into action, enacted legislation, and got something done to fix a problem that does not exist. In the meantime, what other issues are being put on the back burner or completely forgotten about?

These clerical errors, hilariously enough, come back to the SSA's bad death data. If that issue was fixed, the myth of voter fraud would never have had legs to stand on. The SSA's bad death data affects so many different issues, and in this case, it gives credence to a harmful myth that diverts government funding and time from a legitimate issue, like business identity theft, to a fake problem perpetuated by the ill-informed.

I could not wait for Larry to get back from the NASS conference. I had sent him armed with a long list of questions and people to talk to in order to figure out what NASS had been doing to combat business identity fraud since their 2012 report came out. To say the least, it was disappointing to hear how our representatives were spending their time and our money to talk about a fake problem rather than a real one. So what has NASS been doing? Our best investigative powers have not been able to figure out the answer to that.

NASS may not have done much together as an organization in the last few years, but some states have taken it upon themselves to tighten regulations, especially with regard to online business filings. These states have sought to make the process of reinstating a dissolved entity more rigorous, in part because of NASS's 2012 recommendations. At this point, a corporation that was registered in North Dakota can only be reinstated by a court order after one year of being dissolved.[176] That means I cannot log onto the Secretary of State's website and insert myself into the company. I would have to go through the court system, which very few fraudsters would be willing to do. Business owners might be frustrated by the inconvenience, but it is so much better than risking their businesses' identities.

This is where facial recognition technology could really change the landscape of fraud and its capabilities. Facial recognition would help combat boundary blindness by bringing physical likenesses across state lines. This type of inter-agency communication would combat fraud by outlining a stronger form of identity. People have a much harder time changing their faces than their identity cards.

A Face in Fraud

Miguel Roldan was arrested in April 2016. The only problem is that he was not actually Miguel Roldan. A Massachusetts man named Luis Medina assumed Miguel Roldan's identity by getting a copy of his birth certificate and using it to get a driver's license. Medina's fraudulently obtained driver's license was from New Hampshire. If New Hampshire had facial recognition technology in place, they would have been able to screen Medina to see that he was not in fact Miguel Roldan, and that he was actually wanted in Massachusetts for drug trafficking.

As it turns out, Medina had assumed his new identity to escape the charges in Massachusetts. He used his new identity, the identity of Roldan who died in 1994, to get a job with an asbestos company. This company was hired to do work on a naval shipyard in Portsmouth, New Hampshire. Medina was able to get in and out of the shipyard over 30 times. To heighten the matter, this was no ordinary shipyard: it housed a repair shop for nuclear submarines.

Facial recognition technology would have come in handy in this case, except for the fact that states are not allowed to share pictures across state lines. Boundary blindness at its finest. Medina never would have been able to get a license in New Hampshire had facial recognition technology been utilized, and he most likely would have been arrested for his outstanding warrant. Furthermore, it would have prevented him from obtaining access to a naval base. While it seems that Medina was just trying to find stable work, seeing what he fraudulently had access to is alarming.[177]

New Hampshire does not employ facial recognition technology, in part due to privacy laws. This has made New Hampshire a target state, and a significant influx of fraudulent license charges have come in since other states started using facial recognition technologies.[178] A Pew Charitable Trust article notes that at least 39 states employ facial recognition software in some capacity. New York found 14,500 people had two or more licenses because of their facial recognition implementation.

Even in Nebraska, facial recognition technology has uncovered a couple hundred cases.

The significant factor with all those cases is that they occur solely within state lines. New York and New Jersey together ran a test of commercial drivers licenses and found thousands of people with a license in each state. This allowed those with a revoked license in one state to continue earning a living while driving under their second license. Many of them had multiple DUIs and various other offenses.

"If government agencies are not allowed to work with law enforcement, they're really tying our hands, as far as protecting the public," said Betty Johnson, a DMV administrator in Nebraska. "Facial recognition technology is helping protect people not just from identity theft and fraud, but as drivers and as neighbors." We agree. Limiting conversations between government administrators, legislators, and law enforcement—the people who actually implement the government's policies—is absurd. How can anyone find fraud if everyone is just looking within their own agency?

The first-generation fraud tools are not doing enough. Relying on pay and chase methods, and not communicating widely between states, institutions, and agencies, are ineffective methods to reduce fraud. If businesses refuse to address the systemic problem they all face, if they continue to ignore it and write it off as a cost of doing business, nothing will get fixed, and fraud will persist.

Band-aid solutions are no longer enough. Implementing front-end solutions like encryption, facial recognition technology, and limiting data intake will stop the problems at the source. We would be remiss if we wrote a book in 2017 that completely ignored data hacks. Just this week, the credit bureau

Equifax Inc. was hacked, exposing names, Social Security numbers, birth dates, addresses, and driver's licenses. It affected 143 million customers. Credit cards for about 209,000 customers were also accessed.[179] When both Larry and I read the headlines, neither of us were too surprised. These kinds of hacks have become commonplace, and with the total number of exposed identities in the United States greater than the number of citizens, the exposure of identities is no longer shocking.

What is unique to this hack is that Equifax is seen as one of the most secure institutions. When government computers were hacked in 2015, exposing 21.5 million people,[180] the public felt more nervous about it than when retailers were hacked back in 2012 and 2013. As a credit reporting bureau, Equifax holds our most sensitive information, and offers identity protections to boot. If we can't trust them with our sensitive data, who can we trust? Equifax certainly has higher data standards and protections than government agencies and institutions. What is to be done?

Employing front-end solutions is a critical component of preventing any kind of fraud, large-scale hacks included. In a paper discussing hacker innovations and culture (and one of the most fascinating and accessible things I have read on the topic), Sean Zadig and his colleagues examine the hacker-level equivalent of pay and chase. They note that the previous research in their field, while trying to understand hackers themselves, has "not attempted to solve the problem of the constant attacks that organizations face, and as a result the costly cycle of attack and defend has continued unabated."[181]

Zadig discusses how hackers adapt to new technologies put in place to deter them by posting on message boards, so others can learn as well.[182] If businesses and agencies were more like hackers, with half as much communication, they could respond the same way hackers do. Communicating success and failure would help the group as a whole.

Zadig believes his approach of understanding hacker communities, their language and ways of interacting, is key to disrupting their activity. He notes that by infiltrating these types of communities, law enforcement and information security firms would be able to know which individuals to target, so more senior hackers who help distribute information and techniques. He also mentions the power of comment sections, and how "labeling innovations as too simplistic and more appropriate for script kiddies"[183] can be helpful in dissuading other hackers from using those techniques. His paper is fascinating, and gives some insight into the psychological aspects of hacking. While it may seem unrelated, Zadig's unconventional solutions provides some relief. He is applying psychology to these communities and getting underneath the issues, rather than chasing after them. These are front-end solutions.

We hope the recommendations we make can be used across the board—from vital records offices and DMVs, to small businesses to large corporate entities. The connections between identity and fraud are innumerable. Protecting your own identity, your business's identity, and the identities you may be responsible for is paramount to preventing fraud. Front-end solutions and utilizing emergent technologies in efficient ways, and staying ahead of fraudulent trends, are the only ways to effectively combat fraud.

There is so much at stake, that individuals of all political persuasions can agree on the need for new measures. Securing identities protects individuals and businesses from losing assets, it protects benefits payments from going to the wrong individuals,[†] and it protects against people assuming the wrong identity for either financial or criminal reasons. Fraud is the crime of this century. It is endless and endlessly creative. Fraud engages the mind. If we are going to stand a chance against it, we are going to have to use our own.

[†] At this moment in September 2017, both Hurricane Harvey and Hurricane Irma are battering the Caribbean and the southern United States. FEMA relief efforts in Texas are well under way, but if the relief efforts are anything like Hurricane Katrina, there will be significant improper payments made during these first crucial weeks. Unfortunately, we will not know until it is too late.

Epilogue

Most of the recent articles I have read have quipped the all-too tiring cliche that data breaches are becoming the third certainty in life, next to death and taxes. While it may be tired language, it's not wrong. We are fatigued by headlines of all kinds lately, and it makes some people want to opt-out. For many, the shadowy, confusing world of fraud feels far enough away, unrelated to their daily life. The only problem is, when fraud decides to come out from the shadows, it is too late to address the problem.

Perhaps the world we describe in this book just doesn't feel real to you, but trust me, when you're standing at the check-out with your melting bagel bites, and the fraud alert that shut down your card turns out to be a real breach, it will be a terrifying feeling. Fraud is one of those things that dealing with before something goes horribly wrong always seems like such a hassle. Putting a credit freeze on your identity is just so much work. You have to actually call the companies, listen to the horrible option menu, press buttons, listen to some equally horrible elevator music, and eventually talk to an unhelpful person who has to put you on hold to ask their supervisor how to do their job. It's not any fun, but trust me, it is a lot less fun to chase after identity thieves.

It is all terribly depressing to deal with. To look at the Equifax hack, the first major hack of a company built to protect our credit histories. They lost all our data, and then told us it was three times worse than what they originally said. It is hard to trust anyone, any company, and any agency at this point. It is hard for businesses to trust their customers, and for agencies to trust their businesses. And while trust might be the most difficult thing to do, it might be the way to the future. Transparency between all these separate parties may be the one thing fraudsters cannot stand up to. By communicating equally on failures and successes, everyone will come out better for it.

We have explored identity documents and how they are used to breed additional fraudulent documents used by criminals for a variety of reasons. We addressed the emergent frauds of business identity theft, and how it can compound into leveraging individual identities to defraud agencies and commercial entities. The most vexing question is this: how do we begin to change the process and perception of documenting and identifying our citizens in a unique and secure way? With 320 million Americans alive today, and a flawed group of stove-piped systems tracking births, Social Security numbers, marriages, licenses, passports, and deaths, our current methods are unsustainable. Our world encompasses great advances in technology and global commerce, and those changes serve as an excellent breeding ground for fraud.

The impacts of not changing our current system are frightening. The internet is still in its infancy, developing further every day. Criminals are using it effectively to mask their true identities. The United States government currently loses over $100 billion per year to fraud, and this number is increasing at an alarming rate. We can either begin the process of change or wait until it gets worse. The only certainty is that it certainly will not get better on its own.

Our political climate does not help the situation. By pitting those who don't want government to have any additional control against those who think our data should be centralized, it delays decisions, as we saw with the Real ID Act and Vermont's H.111 bill. It creates mistrust of our technologies and fans the flame of unfounded, xenophobic ideologies. Creating front-end technologies that incorporate biometrics, testing, and troubleshooting before implementing them, and actually training employees to understand and properly use them, are how we will fight fraud in the future.

Today there are approximately 15–18 million people impacted by identity theft and fraud every year. Identity theft is not a bacterial infection, it's a virus. Once you have been affected by identity fraud, you have it for life. Already 4.9 million people have received a new Social Security number from the SSA because of continued identity issues. We believe these people deserve some kind of reassurances.

Imagine an identity program a citizen could opt into, but is not required. Think of it like the TSA's PreCheck program: over 3 million Americans have paid $85 to stand in an express line and not take off their shoes. What is the value to protect your identity and not sit on the phone all day to repair your credit, or fight over a bank loan that you never took out? This program could require all credentials to be checked and fingerprints taken just like the TSA PreCheck, but a new national ID card (NIC) would be issued that contains a picture, a full biometrics file on the card, and is chip-enabled just like a credit card. It could also fully replace the Social Security number. The old Social Security number would be frozen for existing commercial and government programs. This number could also serve as your passport number and driver's license number.

What if this system tied into multi-factor authentication, creating an online account in which you incorporate your current cell phone and a unique code. When the card or new Social

Security number is used, it allows government and commercial entities, via this closed system, to contact the phone number and request the code. This is a three factor authentication process, to which other factors can be added for additional security, and all based on an open framework for future capabilities.

On the back end is an entity that stores information but can do real-time transactions. Credit card companies are the masters of this technology, and utilizing their knowledge, and that of other private industry leaders would be crucial to the success of implementing such a large-scale technological enterprise. By incorporating private knowledge with that of government reach, this back-end system could verify with the actual card holder A) a legitimate card, B) that the associated cell phone is being used, C) the cell phone holder knows the code, and possibly D) that the holder of the cell phone has the correct fingerprint matching both the card and the back-end database. Finally, add the capability for a state to upload to this system drivers' license data, putting everything in one place. This type of system could result in real time verification of an identity, minimizing cost and maximizing security.

While all those security measures would help, it is critical to know that there is no perfect system. Anything created today must be as adaptable as the fraudsters they seek to deter. Why would companies and government agencies want to move to a system like this? To cut the multibillion dollar loss per year on both sides, streamline revenue, reduce and speed verification workloads, and increase the delivery of benefits and financial instruments to the proper party.

If you add it up, businesses are losing $221 billion a year, and the U.S. government is losing well over $100 billion a year. Both numbers are increasing by double digits annually. At over a third of a trillion dollars, both business and government have a lot to gain by implementing new solutions and jointly working together on this problem.

It may seem impossible to get private and public institutions to all work together for a common goal, to eliminate unnecessary systems for the greater good, to get American citizens on board with a system that seems Orwellian at best, and it won't be easy. It will require extreme communication and teamwork, but the end goal is something we all hope for. If fraud is allowed in one area, in one type of company, for one individual, it will surely seek a different target eventually.

About the Authors

Larry Benson is currently the Director of Strategic Alliances for the Government Tax and Revenue vertical at LexisNexis® Risk Solutions. In this role, Mr. Benson is responsible for developing partnerships for the tax and revenue vertical. He focuses on embedded companies that have a need for third-party analytics to enhance their current offerings. As principal author of the Fraud of the Day column, Mr. Benson provides commentary on news about fraud against federal, state and local government agencies, and has garnered a following as a national speaker on the topic.

Alana Benson is a researcher, writer, and consultant specializing in document fraud. Ms. Benson has presented to both the Federal Trade Commission and the Fraud Defense Network's advisory board. She has also consulted for publications like *Consumerist* and been featured on the ACFE's 'Fraud Talk' podcast. Ms. Benson previously wrote for LexisNexis's daily fraud column, *Fraud of the Day*, and has written for outfits like Rosen Publishing, BlazeVOX magazine, and ScholarWorks. Ms. Benson holds a BA in English and Classical Studies with a concentration in Ancient Greek Literature.

Acknowledgments

The authors would like to thank, first and foremost, LexisNexis for supporting this project. Thank you to all the contributors, both explicitly stated and anonymous, that contributed to this effort. Without your stories and knowledge, we would have nothing to write. Thanks especially to those who have suffered from identity fraud and know firsthand the trials and tribulations it brings, and who were willing to share their experiences with us. A gracious thanks to Frank Abagnale for once again contributing and making us look good, and Sophie Marsh, an intern who operated like a seasoned veteran. An enormous thank you to our editors, Jenna Pacitto and Julian van der Tak, and all those who read portions and gave feedback. Thank you also to our collective family and friends. Your support made this effort possible. Lastly, thank you to The Middle Fork Restaurant and the Wu-Tang Clan, as it was with your coffee and music that anything got written at all. Thank you as well to Adeline Delores Marie Armour, a girl who is going places.

Endnotes

Note: Due to the dynamic nature of the internet, website links may have changed.

1 Anonymous. "CSI Effect." *LII / Legal Information Institute*, August 19, 2010. https://www.law.cornell.edu/wex/csi_effect.

2 "Kentucky Office of Vital Statistics Agency Information." *VitalChek*. Accessed October 18, 2017. https://www.vitalchek.com.

3 "Arizona Vital Records Agency Information." *VitalChek*. Accessed October 18, 2017. https://www.vitalchek.com.

4 McCoy, Richard Public. "Fraud and Identity Theft Concerns with Vermont's Certified Copies of Birth and Death Certificates." *Vermont Department of Health*, January 15, 2015. P. 9.

5 —"Fraud and Identity Theft Concerns with Vermont's Certified Copies of Birth and Death Certificates." *Vermont Department of Health*, January 15, 2015. P. 10.

6 —"Fraud and Identity Theft Concerns with Vermont's Certified Copies of Birth and Death Certificates." *Vermont Department of Health*, January 15, 2015. P. 15

7 —"Fraud and Identity Theft Concerns with Vermont's Certified Copies of Birth and Death Certificates." *Vermont Department of Health*, January 15, 2015. P. 10

8 —"Fraud and Identity Theft Concerns with Vermont's Certified Copies of Birth and Death Certificates." *Vermont Department of Health*, January 15, 2015. P. 11-12.

9 —"Fraud and Identity Theft Concerns with Vermont's Certified Copies of Birth and Death Certificates." *Vermont Department of Health*, January 15, 2015. P. 11-12.

10 Gibbs Brown, June. "Birth Certificate Fraud." 0EI-07-99-00570. Department of Health and Human Services, Office of Inspector General, September 2000.

11 "Birth Certificate Fraud." 0EI-07-99-00570. Department of Health and Human Services, Office of Inspector General, September 2000. P. 13.

12 Devereux, Dennis J. Act. No. 46 (H.111). Public records; health; Executive Branch; municipal government; judiciary (2017). http://legislature.vermont.gov.
13 Kurkjian, Stephen, Callum Borchers "A Marriage of a Dream and a Scheme." *Northeastern Initiative for Investigative Reporting, republished by BostonGlobe.com.* Accessed August 4, 2017. https://www.bostonglobe.com.
14 Kurkjian, Stephen, and Callum Borchers. "Brattleboro Became A 'wedding Destination' for Green Card Scam." *Northeastern Initiative for Investigative Reporting*, September 19, 2011.
15 Rivard, Danielle. "Woman Accused of Forging Marriage License, Collecting Benefits from Brattleboro Co-Op Victim." *SentinelSource.com*. June 13, 2013. http://www.sentinelsource.com/.
16 "State Secure Identity Practices and Policies in 2015." *The Secure ID Coalition*, 2015. http://www.secureidcoalition.org.
17 Javelin Strategy. "Identity Fraud Hits Record High with 15.4 Million U.S. Victims in 2016, Up 16 Percent According to New Javelin Strategy & Research Study," February 1, 2017. https://www.javelinstrategy.com.
18 "State Secure Identity Practices and Policies in 2015." *The Secure ID Coalition*, 2015. http://www.secureidcoalition.org.
19 CGP Grey. *Social Security Cards Explained*. Accessed April 24, 2017. https://www.youtube.com/.
20 Homeland Security. "Real ID." Official website of the Department of Homeland Security. October 18, 2017. https://www.dhs.gov/real-id.
21 Bovbjerg, Barbara D. "Social Security: Observations on Improving Distribution of Death Information." *United States General Accounting Office*, November 8, 2001. http://www.gao.gov/assets/110/109062.pdf.
22 Pelley, Scott. "Script for 'Dead or Alive.'" CBSnews.com, March 15, 2015. https://www.cbsnews.com.
23 Stone, Gale S. "Numident Death Information Not Included on the Death Master File." *Office of the Inspector General*. September 21, 2016. A-06-16-50069.
24 Bertoni, Daniel. "SSA: Preliminary Observations on the Death Master File." *United States Government Accountability Office*. May 8, 2013. http://www.gao.gov/assets/660/654411.pdf. GAO-13-574T. P. 2
25 —. "SSA: Preliminary Observations on the Death Master File." *United States Government Accountability Office*. May 8, 2013. http://www.gao.gov/assets/660/654411.pdf. GAO-13-574T. P. 2

26 Bovbjerg, Barbara D. "Social Security: Observations on Improving Distribution of Death Information." *United States General Accounting Office*, November 8, 2001. GAO-02-233T. http://www.gao.gov/assets/110/109062.pdf.

27 Social Security Act § 205(r)(3), codified at 42 U.S.C. § 405(r)(3).

28 Larin, Kathryn A. "Program Integrity: Views on the Use of Commercial Data Services to Help Identify Fraud and Improper Payments." *United States Government Accountability Office*. June 2016. GAO-16-624. P. 14.

29 Davis, Beryl H. "Improper Payments: Strategy and Additional Actions Needed to Ensure Agencies Use the Do Not Pay Working System as Intended." *United States Government Accountability Office*. October 2016. GAO-17-15.

30 Bureau of the Fiscal Service. "About Do Not Pay-Improper Payment Initiative." *Do Not Pay* (blog). Accessed October 18, 2017. https://donotpay.treas.gov/about.htm.

31 Bovbjerg, Barbara D. "Social Security: Observations on Improving Distribution of Death Information." *United States General Accounting Office*. November 8, 2001. GAO-02-233T. http://www.gao.gov/assets/110/109062.pdf.

32 Stone, Gale S. "Numident Death Information Not Included on the Death Master File." *Office of the Inspector General*. September 21, 2016. A-06-16-50069. https://oig.ssa.gov.

33 Bertoni, Daniel. "SSA: Preliminary Observations on the Death Master File." *United States Government Accountability Office*. May 8, 2013. http://www.gao.gov/assets/660/654411.pdf. P. 6.

34 — "SSA: Preliminary Observations on the Death Master File." *United States Government Accountability Office*. May 8, 2013. http://www.gao.gov/assets/660/654411.pdf. P. i.

35 — "SSA: Preliminary Observations on the Death Master File." *United States Government Accountability Office*. May 8, 2013. http://www.gao.gov/assets/660/654411.pdf. P. 6-8.

36 — "SSA: Preliminary Observations on the Death Master File." *United States Government Accountability Office*. May 8, 2013. http://www.gao.gov/assets/660/654411.pdf. P. 1-2.

37 Stone, Gale S. "Numident Death Information Not Included on the Death Master File." *Office of the Inspector General*. September 21, 2016. A-06-16-50069. https://oig.ssa.gov.

38 Cristaudo, Frank. "Letter to Mrs. Barbara D. Bovbjerg from Frank Cristaudo.," September 19, 2016.

39 Bertoni, Daniel. "Social Security Death Data: Additional Action Needed to Address Data Errors and Federal Agency Access" The United States Government Accountability Office, November 2013. GAO-14-46. P. 6.

40 Bovbjerg, Barbara D. "Social Security: Observations on Improving Distribution of Death Information." *United States General Accounting Office*. November 8, 2001. GAO-02-233T. http://www.gao.gov/assets/110/109062.pdf.

41 Davis, Beryl H. "Improper Payments: Strategy and Additional Actions Needed to Ensure Agencies Use the Do Not Pay Working System as Intended." *United States Government Accountability Office*. October 2016. P. 14.

42 Stone, Gale S. "Numident Death Information Not Included on the Death Master File." *Office of the Inspector General*. September 21, 2016.

43 Davis, Beryl H. "Improper Payments: Strategy and Additional Actions Needed to Ensure Agencies Use the Do Not Pay Working System as Intended." *United States Government Accountability Office*. October 2016.

44 —"Improper Payments: Strategy and Additional Actions Needed to Ensure Agencies Use the Do Not Pay Working System as Intended." *United States Government Accountability Office*. October 2016.

45 —"Improper Payments: Strategy and Additional Actions Needed to Ensure Agencies Use the Do Not Pay Working System as Intended." *United States Government Accountability Office*. October 2016. P. i.

46 —"Improper Payments: Strategy and Additional Actions Needed to Ensure Agencies Use the Do Not Pay Working System as Intended." *United States Government Accountability Office*. October 2016. P. 6, Footnote 14.

47 Bertoni, Daniel. "Social Security Death Data: Additional Action Needed to Address Data Errors and Federal Agency Access" The United States Government Accountability Office, November 2013. GAO-14-46. P. 14.

48 —"Social Security Death Data: Additional Action Needed to Address Data Errors and Federal Agency Access" The United States Government Accountability Office, November 2013. GAO-14-46. P. i.

49 —"Social Security Death Data: Additional Action Needed to Address Data Errors and Federal Agency Access" The United States Government Accountability Office, November 2013. GAO-14-46. P. 12.

50 Bertoni, Daniel. "Improper Payments: Government-Wide Estimates and Use of Death Data to Help Prevent Payments to Deceased Individuals." United States Government Accountability Office, March 16, 2015. GAO-15-482T. P. i.

51 —"Improper Payments: Government-Wide Estimates and Use of Death Data to Help Prevent Payments to Deceased Individuals." United States Government Accountability Office, March 16, 2015. GAO-15-482T. P. i.

52 —"Improper Payments: Government-Wide Estimates and Use of Death Data to Help Prevent Payments to Deceased Individuals." United States Government Accountability Office, March 16, 2015. GAO-15-482T. P. 11.

53 —"Improper Payments: Government-Wide Estimates and Use of Death Data to Help Prevent Payments to Deceased Individuals." United States Government Accountability Office, March 16, 2015. GAO-15-482T. P.14.

54 Bertoni, Daniel. "Social Security Death Data: Additional Action Needed to Address Data Errors and Federal Agency Access" The United States Government Accountability Office, November 2013. GAO-14-46. P.17-18.

55 Larin, Kathryn A. "Program Integrity: Views on the Use of Commercial Data Services to Help Identify Fraud and Improper Payments." United States Government Accountability Office, June 2016. GAO-16-624. P. 9-10.

56 —"Program Integrity: Views on the Use of Commercial Data Services to Help Identify Fraud and Improper Payments." United States Government Accountability Office, June 2016. GAO-16-624. P. 14.

57 Larin, Kathryn A. "Program Integrity: Views on the Use of Commercial Data Services to Help Identify Fraud and Improper Payments." United States Government Accountability Office, June 2016. GAO-16-624. P. 15.

58 "State Secure Identity Practices and Policies in 2015." The Secure ID Coalition, 2015. http://www.secureidcoalition.org.

59 NAPHSIS. "NAPHSIS Provides Written Testimony to the US Senate Committee on Homeland Security & Governmental Affairs Regarding Electronic Verification of Deaths," March 15, 2017. https://www.naphsis.org.

60 Bertoni, Daniel. "Improper Payments: Government-Wide Estimates and Use of Death Data to Help Prevent Payments to Deceased Individuals." United States Government Accountability Office, March 16, 2015. GAO-15-482T.

61 Trasatti Heim, Rose. "Electronic Death Registration." *National Conference on Health Statistics*. August 2010. https://www.cdc.gov/nchs/ppt/nchs2010/26_trasatti.pdf.

62 "eVital (Electronic Death Registration)." Accessed October 19, 2017. https://www.ssa.gov.

63 Bertoni, Daniel. "Improper Payments: Government-Wide Estimates and Use of Death Data to Help Prevent Payments to Deceased Individuals." United States Government Accountability Office, March 16, 2015. GAO-15-482T.

64 "Electronic Death Reporting System Online Reference Manual: A Resource Guide for Jurisdictions." Westat, Version 1. Figure 1. December, 2016. https://www.cdc.gov.

65 "Electronic Death Registration." *National Home Funeral Alliance*. April 4, 2015. http://homefuneralalliance.org/the-law/electronic-filing/.

66 Stone, Gale S. "Numident Death Information Not Included on the Death Master File." Office of the Inspector General, September 21, 2016. A-06-16-50069.

67 —"Numident Death Information Not Included on the Death Master File." Office of the Inspector General, September 21, 2016. A-06-16-50069.

68 —"Numident Death Information Not Included on the Death Master File." Office of the Inspector General, September 21, 2016. A-06-16-50069.

69 Trasatti Heim, Rose. "Electronic Death Registration." *National Conference on Health Statistics*. August 2010. https://www.cdc.gov/nchs/ppt/nchs2010/26_trasatti.pdf.

70 "Electronic Death Registration System - Vital Records: Mission Statement and Goals." Maine Division of Public Health Systems, October 19, 2017. http://www.maine.gov.

71 Howland, R. E., Li, W., Madsen, A. M., Wong, H., Das, T., Betancourt, F. M., Begier, E. M. Evaluating the Use of an Electronic Death Registration System for Mortality Surveillance During and After Hurricane Sandy: New York City, 2012. *American Journal of Public Health, 105*(11).
72 Mirabootalebi, N., Mahboobi, H., & Khorgoei, T. (2011). Electronic Death Registration System (EDRS) in Iran. *Electronic Physician*.
73 "Electronic Death Registration System (EDRS)." Washington State Department of Health, October 19, 2017. https://www.doh.wa.gov.
74 "Kentucky Electronic Death Registration System (KY-EDRS)." Kentucky Cabinet for Health and Family Services, October 19, 2017. http://chfs.ky.gov.
75 "NAPHSIS | Protecting Personal Identity Promoting Public Health | SYSTEMS." NAPHSIS | Protecting Personal Identity Promoting Public Health. Accessed October 19, 2017. https://www.naphsis.org/systems.
76 "Electronic Death Registry System (EDRS)." Upstate Medical University, State University of New York. Accessed October 19, 2017. http://www.upstate.edu.
77 Stone, Gale S. "Numident Death Information Not Included on the Death Master File." Office of the Inspector General, September 21, 2016. A-06-16-50069.
78 "NAPHSIS | EVVE Fact of Death (FOD)." NAPHSIS | Protecting Personal Identity Promoting Public Health. Accessed April 24, 2017. https://www.naphsis.org.
79 —NAPHSIS | Protecting Personal Identity Promoting Public Health. Accessed April 24, 2017. https://www.naphsis.org/evve-fod.
80 "NAPHSIS | Protecting Personal Identity Promoting Public Health | SYSTEMS." NAPHSIS | Protecting Personal Identity Promoting Public Health. Accessed October 19, 2017. https://www.naphsis.org/systems.
81 —NAPHSIS | Protecting Personal Identity Promoting Public Health. Accessed October 19, 2017. https://www.naphsis.org/systems.
82 "NAPHSIS | EVVE Fact of Death (FOD)." NAPHSIS | Protecting Personal Identity Promoting Public Health. Accessed April 24, 2017. https://www.naphsis.org/evve-fod.
83 "NAPHSIS | Protecting Personal Identity Promoting Public Health | SYSTEMS." NAPHSIS | Protecting Personal Identity Promoting

Public Health. Accessed October 19, 2017. https://www.naphsis.org/systems.

84 NAPHSIS. "NAPHSIS Provides Written Testimony to the US Senate Committee on Homeland Security & Governmental Affairs Regarding Electronic Verification of Deaths," March 15, 2017. https://www.naphsis.org.

85 "NAPHSIS | Protecting Personal Identity Promoting Public Health | SYSTEMS." NAPHSIS | Protecting Personal Identity Promoting Public Health. Accessed October 19, 2017. https://www.naphsis.org/systems.

86 "She Went to Ancestry.com for a Family Tree. Instead, She Found a Stolen Identity." *Washington Post.* Accessed June 14, 2017. https://www.washingtonpost.com.

87 Much of the data from this chapter comes from detailed interviews performed by Sophie Marsh and individual representatives from various state vital records offices.

88 "corporation." *Merriam-Webster.com.* 2017. https://www.merriam-webster.com.

89 Gehrke-White, William E. Gibson, Donna. "Florida Leads Nation in Fraud, ID Theft." *Sun-Sentinel.com.* Accessed September 11, 2017. http://www.sun-sentinel.com.

90 "Enabling Commerce. Mitigating Business Fraud." White Paper, *Dun & Bradstreet.* August 2013. P. 2, Footnote 3.

91 Ibata, David. "ID theft stings Captain D's franchisee," *The Atlanta Journal-Constitution,* March 2, 2012.

92 "Identity Fraud Hits Record High with 15.4 Million U.S. Victims in 2016, Up 16 Percent According to New Javelin Strategy & Research Study." *Javelin Strategy.* February 1, 2017. https://www.javelinstrategy.com.

93 "Form 8-K for Whole Foods Market, Inc." United States Securities and Exchange Commission, June 7, 2017. https://www.sec.gov.

94 "Use of Stolen Business EINs for Tax Fraud." *BusinessIDTheft.org.* 2014. http://businessidtheft.org.

95 "Address Mirroring." *BusinessIDTheft.org.* 2014. http://businessidtheft.org.

96 Bowes, Mark. "Man Accused of Posing as McDonald's Compliance Officer in Multistate Scam Arrested in Henrico." *Richmond Times-Dispatch.* October 14, 2016. http://www.richmond.com.

97 Pitt, David. "Man Charged with False Unemployment Claims in 3 States." Accessed June 28, 2017. http://www.times-standard.com.
98 Experian. "Bust-out Fraud: Knowing What to Look for Can Safeguard the Bottom Line," 2009. https://www.experian.com.
99 —"Bust-out Fraud: Knowing What to Look for Can Safeguard the Bottom Line," 2009. https://www.experian.com.
100 Gross, Daniel. "How to Commit a $200 Million Scam: Inside the Year's Most Shocking Credit Card Fraud." *The Daily Beast*. February 6, 2013. https://www.thedailybeast.com.
101 Dobush, Grace. "How Social Security Numbers Became Skeleton Keys for Fraudsters." *Christian Science Monitor*, November 21, 2016. https://www.csmonitor.com.
102 Colley, Amber. "EIN vs. D-U-N-S Number: How To Benefit From Both," July 16, 2015. http://www.dnb.com.
103 Pankratz, Howard. "Florida Men Indicted on Charges Related to Colorado Business ID Theft." *The Denver Post*. May 31, 2013. http://www.denverpost.com.
104 "Special Report: A Little House of Secrets on the Great Plains." *Reuters*, June 28, 2011. https://www.reuters.com.
105 National Association of Secretaries of State. "Developing State Solutions to Business Identity Theft," January 2012. http://web.mit.edu.
106 Model Bus. Corp. Act § 1.25. Comment 1.
107 National Association of Secretaries of State. "Developing State Solutions to Business Identity Theft," January 2012. http://web.mit.edu/supportthevoter/www/files/2013/12/white-paper-business-id-theft-012612.pdf. P. 8-9.
108 American Bar Association. "2016 Revision to Model Business Corporation Act Makes Its Debut." Accessed October 19, 2017. https://www.americanbar.org.
109 Model Bus. Corp. Act § 1.25. Comment 1.
110 Dirs. Gillman, Terry and Terry Jones. *Monty Python and the Holy Grail: The Screenplay*. UK: Methuen, 2003. P. 76. ISBN 0-413-77394-9.
111 Kent, Lauren. "Where Do the Oldest Americans Live?" *Pew Research Center*, July 9, 2015. http://www.pewresearch.org.
112 Noguchi, Yuki. "Identity Theft A Growing Concern For Businesses." *NPR.org*. Accessed September 14, 2017. http://www.npr.org.

113 Benson, Alana. Interview on AAA Termite & Pest Control and Business Identity Fraud with Scott Burnett. Phone, September 14, 2017.

114 "Revised Refundable Credit Risk Assessments Still Do Not Provide an Accurate Measure of the Risk of Improper Payments." *Treasury Inspector General for Tax Administration.* April 28, 2017. https://www.treasury.gov.

115 —*Treasury Inspector General for Tax Administration.* April 28, 2017. https://www.treasury.gov/tigta/auditreports/2017reports/201740030fr.pdf.

116 "The Paradox of Declining Property Crime Despite Increasing Identity Theft Crime." *Identity Theft Resource Center.* Accessed February 27, 2017. http://www.idtheftcenter.org/images/surveys_studies/ITRConBJSReport.pdf. p. 5.

117 —*Identity Theft Resource Center.* Accessed February 27, 2017. http://www.idtheftcenter.org/images/surveys_studies/ITRConBJSReport.pdf. p. 7.

118 Ford, Matt. "What Caused the Great Crime Decline in the U.S.?" *The Atlantic.* April 15, 2016. https://www.theatlantic.com.

119 Sullivan, Bob. "The Meth Connection to Identity Theft." *Msnbc.com.* March 10, 2004. http://www.nbcnews.com.

120 Cohn, Scott. "Identity Thieves at Their Own Game," *CNBC.* January 31, 2014. https://www.cnbc.com.

121 "Street Gangs Turn to White Collar Crime." Accessed September 10, 2017. *Marketplace.* https://www.marketplace.org.

122 Associated Press. "Street Gangs Migrating from Drugs, Robberies to White-Collar Crime like Credit Card Fraud." *US News & World Report.* Accessed September 10, 2017. https://www.usnews.com.

123 "Global Fraud Report: Vulnerabilities on the Rise." Kroll, 2015. P. 7.

124 Word, Ron. "2.3 Million Consumer Financial Records Stolen." NBCNews.com. July 3, 2007. http://www.nbcnews.com.

125 "Global Fraud Report: Vulnerabilities on the Rise." Kroll, 2015. P. 16.

126 Benson, Larry. "Double Your Crime, Double the Time." *Fraud of the Day,* August 7, 2012. http://www.fraudoftheday.com.

127 Experian. "Bust-out Fraud: Knowing What to Look for Can Safeguard the Bottom Line," 2009. https://www.experian.com/assets/decision-analytics/white-papers/bust-out-fraud-white-paper.pdf.

128 —"Bust-out Fraud: Knowing What to Look for Can Safeguard the Bottom Line," 2009. https://www.experian.com/assets/decision-analytics/white-papers/bust-out-fraud-white-paper.pdf.

129 "Special Report: A Little House of Secrets on the Great Plains." *Reuters*, June 28, 2011. https://www.reuters.com.

130 Marsden, Tom. "Shifting the Fraud Paradigm: From 'Pay and Chase' to Prevention." Dun & Bradstreet, 2011. http://www.dnb.com.

131 "Healthcare Industry Wisdom on Medical Identity Fraud." *Medical Identity Fraud Alliance*. November 2016. http://medidfraud.org/wp-content/uploads/MIFA-Wisdom-Paper-Exec-Summary.pdf.

132 Bertoni, Daniel. "Improper Payments: Government-Wide Estimates and Use of Death Data to Help Prevent Payments to Deceased Individuals." *United States Government Accountability Office*. March 16, 2015. GAO-15-482T.

133 Green, Howard. "Tackle Medicaid's Improper Claims Payments with Data, Not 'Pay and Chase.'" *SyrtisSolutions*. August 15, 2017. http://syrtissolutions.com.

134 Marsden, Tom. "Shifting the Fraud Paradigm: From 'Pay and Chase' to Prevention." *Dun & Bradstreet*. 2011. http://www.dnb.com.

135 Mongan, Emily. "Tackle Medicare Fraud with Data, Not 'Pay-and-Chase,' Lawmakers Urge CMS." *McKnight's*. September 28, 2016. http://www.mcknights.com.

136 Dhawan, Vikram, and Carlos Garcia-Pavia. "What Financial Services and Insurers Can Teach Each Other." *Journal of Insurance Fraud in America*, Coalition Against Insurance Fraud, 4, no. 3 (Spring 2015): P. 9–13.

137 —"What Financial Services and Insurers Can Teach Each Other." *Journal of Insurance Fraud in America*, Coalition Against Insurance Fraud, 4, no. 3 (Spring 2015): P. 9–13.

138 Davis, Jessica. "Frankenstein Approach to Cybersecurity Renders Healthcare Organizations Dead Last at Fixing Vulnerabilities The Biggest Problem? Outdated IT and Tight Budgets Make It Hard to Fend off Growing Throngs of Hackers." *HealthcareITNews.com*. October 19, 2016. <http://www.healthcareitnews.com/author/jessica-davis>.

139 —"Frankenstein Approach to Cybersecurity Renders Healthcare Organizations Dead Last at Fixing Vulnerabilities The Biggest Problem? Outdated IT and Tight Budgets Make It Hard to Fend off

Growing Throngs of Hackers." *HealthcareITNews.com*. October 19, 2016. <http://www.healthcareitnews.com/author/jessica-davis>.

140 —"Frankenstein Approach to Cybersecurity Renders Healthcare Organizations Dead Last at Fixing Vulnerabilities The Biggest Problem? Outdated IT and Tight Budgets Make It Hard to Fend off Growing Throngs of Hackers." *HealthcareITNews.com*. October 19, 2016. <http://www.healthcareitnews.com/author/jessica-davis>.

141 National Association of Secretaries of State. "Developing State Solutions to Business Identity Theft," January 2012. P. 9.

142 —"Developing State Solutions to Business Identity Theft," January 2012.

143 —"Developing State Solutions to Business Identity Theft," January 2012.

144 Vijayan, Jaikumar. "Colorado Warns of Major Corporate ID Theft Scam." *Computerworld*. July 16, 2010. http://www.computerworld.com.

145 —"Colorado Warns of Major Corporate ID Theft Scam." *Computerworld*. July 16, 2010. http://www.computerworld.com.

146 National Association of Secretaries of State. "Developing State Solutions to Business Identity Theft," January 2012. P. 11.

147 — "Developing State Solutions to Business Identity Theft," January 2012. P. 12.

148 Benson, Larry. "Interview with Wayne Williams, and Gary Zimmerman, Chief of Staff, Colorado Secretary of State." March 22, 2017.

149 National Association of Secretaries of State. "Developing State Solutions to Business Identity Theft," January 2012. P. 14.

150 Benson, Larry. "Interview with Ralph Gagliardi, Investigations Agent, Alberta Bennett, Operations Manager and Business Licensing, Trevor Timmons, CIO, and Mike Hardin, Director, Business and Licensing. Colorado Secretary of State Staff." March 22, 2017.

151 National Association of Secretaries of State. "Developing State Solutions to Business Identity Theft," January 2012. P. 17.

152 "Identity Fraud Hits Record High with 15.4 Million U.S. Victims in 2016, Up 16 Percent According to New Javelin Strategy & Research Study." *Javelin Strategy*. February 1, 2017. https://www.javelinstrategy.com.

153 Finklea, Kristin. "Identity Theft: Trends and Issues." *Congressional Research Service*. January 16, 2014. https://fas.org/sgp/crs/misc/R40599.pdf. P. 6.

154 Kassner, Michael. "Business ID Theft: Slow Progress in the Battle against Fraudsters." *TechRepublic*. May 19, 2016. http://www.techrepublic.com.

155 —"Business ID Theft: Slow Progress in the Battle against Fraudsters." *TechRepublic*. May 19, 2016. http://www.techrepublic.com.

156 United States Federal Advisory Committee on False Identification. *The Criminal Use of False Identification: a Summary Report On the Nature, Scope, And Impact of False ID Use In the United States, With Recommendations to Combat the Problem*. Washington: U.S. Dept. of Justice, 1976.

157 Davis, Julie Hirschfeld. "Hacking of Government Computers Exposed 21.5 Million People." *The New York Times*. July 9, 2015, sec. U.S. https://www.nytimes.com.

158 "Puerto Rico Birth Certificates Law 191 of 2009." Puerto Rico Vital Statistics Record Office, 2009. P. 1.

159 "Those with Fake Birth Certificates Find It's Easy to Live the American Dream." *Watchdog.org*, June 23, 2014. http://watchdog.org.

160 "Puerto Rican Birth Certificates Issued Before July 1, 2010 Declared Void." *American Civil Liberties Union*. Accessed September 11, 2017. https://www.aclu.org.

161 Chardy, Alfonso. "Puerto Rican Birth Certificates Are Sold to Undocumented Immigrants." *Miami Herald Online*. April 19, 2015. http://www.miamiherald.com.

162 —"Puerto Rican Birth Certificates Are Sold to Undocumented Immigrants." *Miami Herald Online*, April 19, 2015. http://www.miamiherald.com.

163 "Those with Fake Birth Certificates Find It's Easy to Live the American Dream." *Watchdog.org*, June 23, 2014. http://watchdog.org.

164 "Birth Certificates: Recommendations for Birth Certificate Security." *The Document Security Alliance*. May 15, 2015. http://www.documentsecurityalliance.org/forms/Birth_Certificate_Paper.pdf.

165 Kossman, Sienna. "8 FAQs about EMV Credit Cards." *CreditCards.com*. Accessed September 6, 2017. http://www.creditcards.com.

166 Tepper, Taylor. "Why Your Credit Card Now Has a Chip (& Why You Should Care)." *Money*. Accessed September 5, 2017. http://time.com.

167 Paul Starr and Sandra Starr, "Reinventing Vital Statistics: The Impact of Changes in Information Technology, Welfare Policy, and Health Care," *Public Health Reports,* 110 (September/October 1995): 534-544.
168 Fuson, Larry. "Dna enabled certificate. US20110316267 A1." Patent filed June 23, 2011, and issued December 29, 2011. http://www.google.com/patents/US20110316267.
169 Schweikert, Christina, Mark F. Tannian, and Ying Liu. "Securing Birth Certificate Documents with DNA Profiles." St. John's University, 2017. doi: 10.24251/HICSS.2017.290.
170 Glaser, April. "Biometrics Are Coming, Along With Serious Security Concerns." *WIRED*. Accessed September 28, 2017. https://www.wired.com.
171 —"Biometrics Are Coming, Along With Serious Security Concerns." *WIRED*. Accessed September 28, 2017. https://www.wired.com.
172 —"Biometrics Are Coming, Along With Serious Security Concerns." *WIRED*. Accessed September 28, 2017. https://www.wired.com.
173 Kassner, Michael. "AI Stops Identity Fraud before It Occurs." *TechRepublic*. January 14, 2016. http://www.techrepublic.com.
174 Hawes, Katherine. "Identity Theft Costs Businesses $221B Per Year... What Can You Do to Protect Your Operation?" *Dynamic Business*. January 10, 2017, sec. Security, Technology. http://www.dynamicbusiness.com.
175 "Trump's Bogus Voter Fraud Claims." *FactCheck.org*. October 19, 2016. http://www.factcheck.org.
176 N.D. Cent. Code § 10-19.1-148
177 Rothstein, By Kathy Curran and Kevin. "Identity Imposter Gained Access to Navy Shipyard under Assumed ID." *WCVB*. May 2, 2017. http://www.wcvb.com.
178 McElveen, Josh. "State Police Say Facial Recognition Could Help Prevent Driver's License Fraud." *WMUR*. February 23, 2017. http://www.wmur.com.
179 Womack, Brian. "Equifax Says Cyberattack May Have Hit 143 Million Customers." *Bloomberg.com*. September 7, 2017. https://www.bloomberg.com.
180 Davis, Julie Hirschfeld. "Hacking of Government Computers Exposed 21.5 Million People." *The New York Times*. July 9, 2015, sec. U.S. https://www.nytimes.com.

181 Zadig, Sean M. *Understanding the Impact of Hacker Innovation upon IS Security Countermeasures.* Doctoral dissertation. Nova Southeastern University. Retrieved from NSUWorks, College of Engineering and Computing. (976), 2016. http://nsuworks.nova.edu/gscis_etd/976.

182 —*Understanding the Impact of Hacker Innovation upon IS Security Countermeasures.* Doctoral dissertation. Nova Southeastern University. Retrieved from NSUWorks, College of Engineering and Computing. (976), 2016. http://nsuworks.nova.edu/gscis_etd/976. P. 82.

183 —*Understanding the Impact of Hacker Innovation upon IS Security Countermeasures.* Doctoral dissertation. Nova Southeastern University. Retrieved from NSUWorks, College of Engineering and Computing. (976), 2016. http://nsuworks.nova.edu/gscis_etd/976. P. 205.